HUBBLE'S UNIVERSE

HUBBLE'S UNIVERSE

A NEW PICTURE OF SPACE

Simon Goodwin

with a Preface by
JOHN GRIBBIN

CONSTABLE · LONDON

First published in Great Britain 1996 by Constable and Company Limited
3 The Lanchesters, 162 Fulham Palace Road, London W6 9ER
Text copyright © Simon Goodwin 1996. Introduction copyright © John Gribbin 1996
ISBN 0 09 476330 5
Set in Monophoto Sabon by Servis Filmsetting Ltd, Manchester
Printed in Hong Kong through World Print Ltd

A CIP catalogue record of this book is available from the British Library

FOR MUM AND DAD

CONTENTS

Author's Note

This book follows standard scientific practice in using the (American) billion to mean a thousand million (1,000,000,000) not the (British) billion of a million million (1,000,000,000,000).

PREFACE

John Gribbin

If you were on an isolated mountain top, far from any artificial lights, on a completely cloudless, moonless night, you would be able to see no more than a couple of thousand individual stars with the unaided human eye. You would also see a prominent band of light, like a glowing cloud in the sky, the Milky Way. The Milky Way is composed of hundreds of billions of stars, too faint to be distinguished by the human eye, which make up a disc-shaped 'island in space' known as the Galaxy. But those stars are faint only because of their distance from us; they are each typically about as bright as our Sun, or brighter. We see the Galaxy from the inside, where our Sun is just one unspectacular star among those hundreds of billions, located closer to the edge of the disc than to its centre.

Until the 1920s – i.e. until a generation ago (taking the biblical three score years and ten as the span of a human life) – it seemed that that was all there was to the Universe. 'Only' a few hundred billion stars, in a disc-shaped island about 100 000 light years across. This is still a huge collection of stars by any human standard of measurement. A light year is the distance that light, travelling at 300 000 k.p.s., covers in one year – an impressive 10 000 billion km, in round terms. But in the 1920s, as new telescopes became operational, astronomers discovered that what they had previously seen only as fuzzy blobs of light in the sky are in fact other galaxies in their own right, other islands in space, far beyond the Milky Way. The whole of our Galaxy was suddenly diminished in importance from being thought of as the entire Universe to being just one dust mote floating in a vastly greater sea, no more significant among the galaxies than our Sun is among the stars.

One of the leading lights in this dramatic change in astronomers' image of the Universe was the American observer Edwin Hubble, who showed not only that the Universe is much bigger than previous generations had thought, but that it is expanding, with galaxies moving further apart from one another as time passes. This discovery led to the realisation that the Universe must have been born at a definite moment, when everything that we can see today emerged from a tiny, hot, superdense fireball, the Big Bang. It was in Hubble's honour that the Hubble Space Telescope, which obtained the breathtaking pictures of the Universe presented in this book, was named.

The Hubble Space Telescope (HST) isn't just about probing even deeper into the Universe than Hubble himself, or his successors using ground-based telescopes, could probe. From above the obscuring layer of the Earth's atmosphere, the telescope has brought us unprecedentedly clear images of objects in our own astronomical backyard, the Solar System. It has also shown us clear pictures of activity in the interstellar medium of our own Galaxy, where stars are being born, and the debris from stars that have exploded at the ends of their lives. Beyond the Milky Way, the HST has shown us how galaxies themselves interact with one another, swallowing each other up in mergers that create even bigger star systems. And it has revealed the presence of black holes, each with millions of times as much mass as our Sun, devouring stars whole in the hearts of some of those galaxies. But it is in addressing the big questions of just how big the Universe really is, how old it is, and whether it will expand for ever or will itself one day, in the far distant future, cease expanding and collapse back into a Big Crunch, mirroring the Big Bang in which it was born, that the HST really comes into its own.

One of the peculiarities of observations on these very large scales – both a disadvantage and an advantage for astronomers – is that we see distant objects as they were long ago, when the light now reaching our telescopes left those objects. Because light travels at a finite speed, it takes a finite time to cross space, and we see a galaxy that is, say, 10 million light years away as it was 10 million years ago, not as it is today. The problem is that this means astronomers have to be careful about how they compare objects at different distances across the Universe. One galaxy may look different from another simply because it is a different age. In the biological world, you would never guess that a tadpole and an adult frog were the same kind of creature if you only saw young tadpoles and fully mature frogs. But the advantage is that by comparing enough galaxies at many different distances across the Universe (and therefore different ages), it should be possible to get an idea of how galaxies evolve, just as you would quickly realise the relationship between tadpoles and frogs if you had examples of all the stages in the creature's life cycle to examine. The unique power of the HST makes this kind of statistical study easier (or, at least, less difficult) than ever before; some of the images it provides come from objects so distant that we see them as they were when the Universe was only 10% of its present age – we are looking back 90% of the way to the Big Bang itself.

But how can we get some kind of grasp on just how big the visible Universe is? Even the nearest large galaxy to our own, the Andromeda Galaxy, is more than 2 million light years away from us, and is seen by light which has spent more than 2 million years on its journey across space to the Milky Way. Even before the era of the HST, astronomers had measured distances to galaxies out to 1000 million light years, a volume of space encompassing about 100 million galaxies, although very few of them have yet been studied in detail.

One way to picture all of this is to bring everything down to a more homely scale. First, imagine the Sun, which is a typical star, to be reduced to the size of an aspirin. On that scale, the nearest star would be represented by another aspirin some 150 km away. this is typical of the separation between stars (apart from stars which occur in binary or more complex systems, held in orbit around each other by gravity.

Now change the scale so that the entire Milky Way Galaxy is represented by an aspirin. Just as stars sometimes occur in groups held together by gravity, so do galaxies. Our near neighbour (by cosmological standards) the Andromeda Galaxy is, in fact, just such a companion to our Milky Way Galaxy, and on the scale in which an aspirin represents the Milky Way, the Andromeda Galaxy would be represented by another aspirin just 13 cm away. But even the distances between groups of galaxies are, relatively speaking, much less than the distances between stars. The nearest equivalent small group of galaxies to our own Local Group is called the Sculptor Group, and on the scale in which an aspirin represents a galaxy the Sculptor Group would be only 60 cm away. And only 3 m away we would find a huge group of about 200 aspirins (each one representing a galaxy containing hundreds of billions of stars) in a swarm about as big as a basketball. Some other, more distant, clusters of galaxies would each fill a sphere 20 m across, on this scale. But the entire visible Universe would be represented by a sphere only about 1 km across.

Galaxies are very much closer together, relatively speaking, than the spacing between stars within a galaxy. This is just as well for cosmologists – if the separation between galaxies were as great, in proportion, as the separation between stars, the distance to our nearest neighbouring galaxy would be about a hundred times greater than we have yet probed with Earth-based telescopes. No astronomer would yet have detected any galaxy beyond the Milky Way, and we would still think that our own Galaxy was the entire Universe.

But how do astronomers measure such vast distances at all? The key lies in the existence of a class of stars known as Cepheid variables, which change in brightness regularly, first brightening, then fading, then brightening up again. It turns out that the average brightness of an individual Cepheid depends on the length of its cycle of variation – the time it takes to go from maximum brightness through its dim phase and back to maximum brightness. So by measuring the period of the variation of a Cepheid, it is possible to work out how bright it really is. Then, by measuring its apparent brightness on the sky, astronomers can infer its distance – the fainter the star seems to be, the further away it must be. The distances to just a handful of Cepheids within our Galaxy have been measured by other, more laborious techniques, which I won't explain here. What matters is that using this calibration of the Cepheid distance scale, if astronomers can measure the periods of Cepheids in other galaxies, they can determine the distances to those galaxies.

This was the technique used by Hubble to measure distances to a few nearby galaxies. Hubble also discovered that the light from other galaxies has been altered in a particular way. When the light from a star or galaxy is split up by a prism to form a rainbow pattern, the rainbow colours in the resulting spectrum are crossed by series of bright and dark lines. These lines are a kind of fingerprint, characteristic of the atoms that are producing the light – the familiar orange-yellow of many streetlights, for example, is produced by two bright lines in the spectrum associated with sodium atoms. Each set of lines occurs at a characteristic set of wavelengths. But in light from distant galaxies, the lines are shifted bodily towards the red end of the spectrum. This is the famous redshift, produced by the expansion of the Universe. Hubble discovered that the redshift is proportional to distance. So if the redshift/distance scale can be calibrated, the distance to even the most remote objects we can see with our telescopes can be inferred simply by measuring their redshifts.

The problem is to calibrate the redshift/distance relation. One of the best ways to do this would be to identify individual Cepheid variables in galaxies in clusters far beyond our Local Group. That is extremely difficult using ground-based telescopes, so until recently there has been considerable uncertainty in the cosmic distance scale. But the HST is now carrying out a series of observations of Cepheids in distant galaxies, observations which should soon reduce these uncertainties, and fix the distance scale of the Universe. And once we know exactly how far apart the galaxies are today, as well as how fast they are separating, it will be a relatively simple matter to work backwards to determine exactly how long ago they began this expansion in the Big Bang. At present, all we can say is that the Big Bang occurred somewhere between 10 billion and 20 billion years ago, with early results from the HST favouring the younger end of the range.

The same information that will pin down the date of birth of the Universe will enable astronomers to determine whether the present expansion will ever end – knowing how far apart the galaxies are, and how fast they are moving, will make it possible to work out when, if ever, gravity will bring the expansion to a halt. So the HST really is about to answer the ultimate questions concerning the origin of the Universe and its ultimate fate.

That, of course, is the scientific justification for the telescope. But you don't have to know anything about science, or be unduly interested in the origin and fate of the Universe, in order to enjoy the spectacular images from the telescope, some of the best of which have been collected in this book. Above all else, the HST has brought back to centre stage the real reason why anybody does science in the first place – the sense of wonder at the nature of the world which we inhabit. That was certainly why I became a scientist, and the images gathered here have given me an opportunity to rekindle that sense of wonder. I hope they stir something in you.

INTRODUCTION

Astronomy is the oldest science. From the dawn of Man we have looked up into the night sky and seen the stars far above us. Our ancestors saw how the appearance of the heavens was ordered and predictable. The Moon and the bright planets moved against an apparently fixed backdrop. Although the stars did not change position relative to one another, the stars visible at night changed with the seasons. People realised that the passage of time or their position on the Earth could be calculated by observing the positions of the stars. Such practicalities came hand-in-hand with an unquenchable curiosity about what was around us in the Universe.

At first it was assumed that the Earth was the centre of the Universe and Man the most important thing within it. Though some in ancient Greece thought that the Sun might be the centre, it was not until Copernicus (1473–1543) proposed that the Earth did not hold a prime position in the scheme of things that general perceptions of our position in the Universe began to change. The telescope was one of the most vital inventions to further this understanding.

What we can see of the sky with our unaided eye is only a tiny fraction of what the Universe contains. The quest of astronomers from the earliest times has been to find a way to see more and more of the Universe, and our knowledge of it is inextricably linked with the development of the telescope – our eyes on the Universe. The better the telescope, the more we can know, and the deployment of the Hubble Space Telescope is increasing our knowledge by leaps and bounds.

Early telescopes

One of the first to study the sky in detail through a telescope was the Italian Galileo Galilei (1564–1642). He built a refracting telescope in June 1609 which was far better than any other to date. He showed his invention to the Italian military and was offered a life-time tenure to work at Padua. Later that year, Galileo's observations of the Moon showed that it was not a smooth sphere as the Greeks had thought, but rough, with mountains and valleys like those on Earth. The following year, Galileo discovered that Jupiter had four moons that circled it in a 'mini Solar System', and that Venus had phases just like the Moon, so that it too must be circling the Sun.

Though these ideas did not find favour with the Church (indeed, it was not until 1979 that the Vatican finally officially conceded that Galileo was right!), the modern science of astronomy was born, and from that day to this scientists have been working to improve the power of telescopes. By the 1660s, larger refracting telescopes were being constructed but they were very unwieldly. The technology of the time could not overcome the problem of 'chromatic aberration' – the tendency of a lens to split the light into its component colours, which gave images coloured fringes – but it could at least be reducecd by using thin lenses of very long focal length. But then the eyepiece had to be a long way from the main lens, and the resulting telescopes were long, flimsy objects supported by masts and ropes. Then, in 1672, Isaac Newton (1643–1727) at a meeting of the Royal Society in London exhibited his reflecting telescope. This used a mirror rather than a lens to focus the light. The advantage of this system is that mirrors, as well as avoiding the problem of coloured fringes, can be built larger and more accurately than can lenses, with the result that the power of the telescope increased and more of the Universe began to open up to scrutiny. It was not until the twentieth century, however, that the true scale of the Universe and the insignificance of Earth and our Solar System began to be clear. The Milky Way, in which the Sun is one of hundreds of billion stars, is only one of billions of galaxies in the Universe.

Hubble

The notion of the 'expanding Universe' is inextricably associated with the American astronomer Edwin Powell Hubble (1889–1953), a pioneer in determining the existence of galaxies other than the Milky Way. It is therefore entirely appropriate that in 1983 the decision was taken to name in his honour the project for the largest observatory in space so far. Having studied physics and astronomy at the University of Chicago (where he also shone as a heavyweight boxer), Hubble took a law degree at Oxford and practised briefly as a lawyer; he also served in the First World War. He then returned to Chicago for his PhD in astronomy, and spent most of his career at the Mount Wilson Observatory in California.

The possibility of an orbiting telescope on board a manned space station was first raised in 1923 by the German rocket pioneer Hermann Oberth. Placing a telescope above the Earth's atmosphere would remove many problems. The atmosphere prevents most of the electromagnetic spectrum from reaching the Earth's surface as well as degrading the quality of that which does. Such forward thinking was regarded with scepticism at a time when air travel was still a novelty. His idea was resurrected in 1946 by the American astronomer Lyman Spitzer Jnr, but not until the early 1960s did it begin to be taken seriously. Space travel was advancing in leaps and bounds: the first Sputnik had been launched in 1957; the first manned spacecraft took off, also from Russia, in 1961; the *Apollo 11* mission to the Moon, a triumph for America's National Aeronautical and Space Administration (NASA) which culminated in the now proverbial 'giant leap for mankind', was in 1969. With this background anything seemed possible, and plans began to be drawn up for a Large Space Telescope.

The proposal that was eventually to become the Hubble Space Telescope (HST) was put forward in 1977, with an estimated cost of $450 million and a launch date set for 1983. However, NASA was told that it had to find a partner to help spread the costs. NASA approached the European Space Agency (ESA), which agreed to provide 15% of the money required (in the form of components for the HST)in return for 15% of the observing time on the telescope. This portion of the money was to be provided by ESA developing and building various components of the telescope.

The enormous technological problems encountered in building a satellite such as the HST resulted in considerable delays. A new date was set for late 1986, but the tragedy of the *Challenger* disaster and the subsequent grounding of the entire shuttle fleet meant that, once again, the launch of the HST was postponed, this time until 1990. Meanwhile the costs of the project were escalating: the estimate for designing and building the telescope had more than trebled to $1.6 billion by 1986. The news was not all bad, however. Engineers used the time during the enforced delay to test and further improve many of the HST's systems. The Space Telescope Science Institute (STScI) was established at Johns Hopkins University in Baltimore, and has become the centre for the administration and science of the HST.

The aims of the HST

The HST was designed to make a large number of very different observations over the entire range of astronomical objects, from asteroids and comets in the Solar System to supermassive clusters of galaxies at the edge of the Universe. The differences are staggering – a big asteroid might be some 500 m across, a super galaxy cluster may be a billion billion times larger. While the telescope cannot get the detailed images of other planets in the Solar System that such probes as *Voyager* and *Mariner* have, it has the advantage of being able to make more frequent and sustained observations, looking for changes in the atmosphere and temperature. It can also seek out tiny asteroids and comets that would be invisible even to the best ground-based telescopes. The HST came into its own when it was used to watch the comet – Jupiter collision in July 1994 (see plates 14 and 15).

The HST is also being used to probe our own Milky Way. Its superior resolution enables it to see the details of star formation in huge clouds of gas and dust. Other galaxies can be seen in great detail, and their structures and components can be studied.

The HST is helping astronomers to determine the crucial 'distance ladder'. The distances to astronomical objects outside our own Solar System cannot be measured directly, so indirect and often uncertain methods must be used. The further away an object is, the more inaccurate our estimate of its distance normally is. Distances are usually cited in terms of 'astronomical units' (the average distance between the Earth and the Sun – 150 million kilometres) or in 'light years' (the distance that light travels in a year – 9500 billion kilometres). When distances to stars are quoted they often have an error of 5 or 10%; for galaxy clusters this error can be higher than 50%. With the HST's improved resolution, stars and galaxies further away than before can have their distances from Earth accurately determined.

In Hubble's expanding Universe, the HST is being used to find out more about the rate of expansion. On the basic premise of Hubble's law, that the further away a galaxy is, the faster it is moving away from us, cosmologists are now attempting to determine how fast this expansion is, i.e. to measure the 'Hubble constant'. Estimates for the Hubble constant vary by a factor of about two, a discrepancy which it is one of the HST's main aims to try to reduce.

The most important and exciting work that the HST is engaged in is attempting to determine the size and nature of the entire Universe. Cosmologists think that the Universe was formed from a tiny, superdense, superhot point some 10 to 20 billion years ago in the Big Bang. The Universe has been expanding out from that point, growing larger and larger until it has reached the size it is today. Will the Universe continue to expand, or will it eventually recollapse? How fast is this expansion? How old is the Universe? How big is the Universe? The HST is playing a vital role in helping us to find answers to these questions.

Hubble's Instrumentation

The HST is a large satellite, a cylinder 13.1 m long and 4.3 m in diameter at its widest point, weighing 11.6 tonnes – about the size of a single-decker bus. The HST was designed to do a very wide variety of astronomical jobs and it has to do these in the vacuum of space, far from the helping hands of its human operators. It therefore requires some of the most technically complex astronomical instrumentation yet built (see plate 1).

As it is in orbit the HST is able to observe, unhindered by atmospheric interference of any kind, all frequencies from the infrared, through the entire visible spectrum and into the ultraviolet. Light enters the tube of the telescope and bounces off the main 2.4 m diameter mirror and is focussed towards a smaller mirror above. From this mirror the light is reflected through a small hole in the centre of the main mirror and into the telescope's instruments (an arrangement of mirrors used in many ground-based telescopes as well). This mirror would enable the HST to resolve the disc of a five-pence piece from a distance of over 20 km, and to see the light of a firefly from about 16 000 km.

The HST's instruments are grouped into two sets. At the core of the telescope are the 'on-axis' instruments that are able to observe objects towards which the telescope is pointing. Around these, in the four radial bays, are the 'off-axis' instruments, which can see a slightly larger area. One of these radial bays contains the Wide Field and Planetary Cameras (collectively the WF/PC or wiff-pic), the most frequently used equipment on the spacecraft, which took most of the pictures in this book. Built at NASA's Jet Propulsion Laboratory under the direction of its principal investigator James Westphal of the California Institute of Technology, they are sensitive to frequencies from the infrared to the ultraviolet, a range far wider than that of which the human eye is capable.

The WF/PC receives the light from the telescope in 'charge coupled devices' (CCDs),

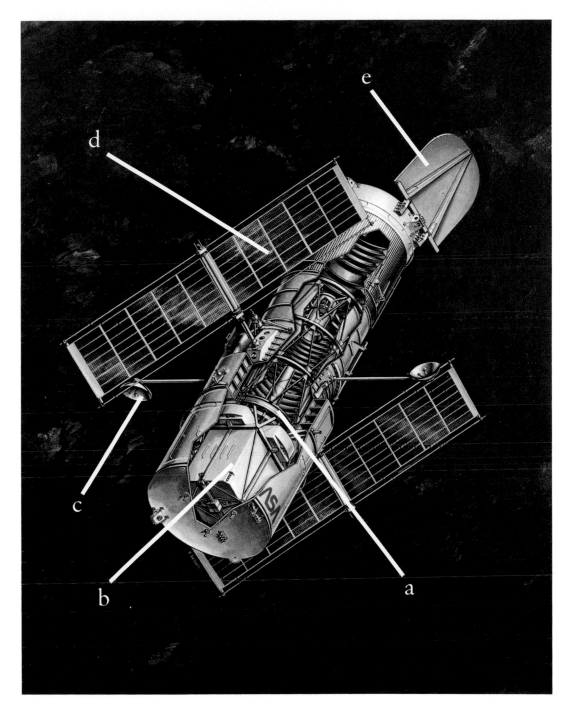

PLATE 1

(a) Main mirror. (b) Instrumentation bay. (c) High-gain antenna (communications equipment).
(d) Solar panel. (e) 'Sun screen'.

which are electronic forms of photographic plates that store the information about the light that has hit them in digital form. A CCD is made up of a large array of tiny electronic 'eyes' (pixels) that turn the light that they receive into an electrical signal. The signals thus produced are stored on a computer and combined to generate an electronic image, rather like the way in which the chemicals in a photographic emulsion are (permanently) changed by the light that falls upon them and builds up a photographic image.

The other three radial bays contain the Fine Guidance Sensors (FGSs). These instruments are used to ensure that the telescope points in the correct direction. They achieve this by finding and locking onto 'guide stars', bright stars whose position is known to a high degree of accuracy. Only two FGSs are required to orient the telescope, but they perform such a vital function that a 'spare' was considered essential.

In the heart of the telescope are the on-axis instruments. The Faint Object Camera (FOC) was designed and built by ESA as one of Europe's contributions to the Hubble project, with H.D. van der Hulst of Leiden Observatory in the Netherlands heading the design team. The FOC utilises the incredible resolving power of the HST to its full. Its image-intensifying techniques enable astronomers to observe objects that are 50 times too faint to see from the Earth's surface. The FOC can observe to a resolution of two-hundredths of an arcsecond.

In the axial bays there are also two spectrographs, the first of which is the Faint Object Spectrograph (FOS), built by a team under Richard Harms of Science Applications Inc. The FOS has a wide range of frequencies that it can observe (and so a larger number of signature lines in a spectrum that it can see). The FOS was a slight casualty of the launch delay caused by the *Challenger* disaster in that one of the mirrors was damaged during the four years of storage. The mirror oxidised slightly, so that it does not reflect as well as it should, especially in the ultraviolet region of the spectrum. The second spectograph is the Goddard High Resolution Spectrograph (GHRS), with John C. Brandt its principal investigator. It is an ultraviolet spectrograph, designed to look at some parts of the spectrum that the atmosphere makes inaccessible to us on the ground. The GHRS selects a small portion of the spectrum and can look at the structure in very fine detail.

A further instrument, the High Speed Photometer (HSP), was installed in the axial bays. The HSP was designed to look at particular frequencies and provide detailed information about how the light from a particular source changes with time. Unfortunately the HSP was not to last, and had to be replaced in the repair mission (see below).

Observing with the HST

The HST observes over the entire sky during the course of a year. It cannot observe objects too close to the Sun (within 50°) because of the damage that would be caused by sunlight entering the telescope. For the same reason the HST cannot observe too close to the Earth or Moon. Its only other restriction is due to the South Atlantic Anomaly: over the South

Atlantic there is a region of highly energetic, charged particles collected by warps in the Earth's magnetic field; these particles interfere with spacecraft electronics, so observations over this region will not be reliable. Observations are therefore often scheduled so that the HST will be moved so as to aim towards a new target during passage above this region.

The STScI plans the HST's observing programme, making the scientific decisions about how the satellite would best be used. Once this has been decided, the observing schedules are sent to NASA's Goddard Space Flight Center Space Telescope Operations Control Center (STOCC), from where the HST is controlled and its observations are monitored. The STOCC also supervises 'housekeeping' operations of the telescope, which include communications, downloading data to Earth and switching observing targets. The HST can continuously observe targets that are in the directions of the Earth's poles, but other areas of the sky are occasionally eclipsed by the Earth or Moon. When a target is out of sight the HST may change targets, but more normally it will use this time for housekeeping.

The HST has six gyroscopes, of which it requires only four to manoeuvre. These can rotate the satellite completely in about an hour, although observations are normally ordered in such a way that the amount of time spent rotating between objects is minimised. After a rotation a further 75 minutes may be required for the FGSs to find their guide stars. Once the HST has found what it is looking for, it points its mirror directly at the object and the astronomical observations can begin. The light from the source is focussed into the instruments and onto their CCDs or spectrographs. The important difference for an astronomer between the human eye and a CCD is that the latter can 'integrate'. The human eye 'resets' itself many times a second, that is, it wipes clean what it has just seen and begins to observe again, thus enabling us to perceive movement. However, the amount of light that reaches us from most astronomical objects is so small that the human eye, in each fraction of a second that it collects the light, is unable to distinguish it and so the object is 'invisible'. A telescope also increases the area onto which the light falls (this is one of the reasons why the size of a telescope's mirror is so important), so there is more of it for us to see. The invention of the photographic plate (and its successor, the CCD) greatly enhanced this capacity. By contrast with holiday snapshots, for which a camera shutter opens for only a few hundredths of a second, images of faint astronomical objects require the 'shutter' to be kept open for a very long time, sometimes a few hours, before enough light has fallen onto the CCD. All the images in this book are the result of long exposure times.

While the HST is observing, it relays information to White Sands, New Mexico, the communications centre to both STOCC and STScI, via the Tracking Data Relay Satellite (TDRS). When the HST is out of sight of TDRS it can store information on two tapes on board. The information on these tapes can then be relayed to Earth when TDRS is back in view.

It is very rare for an astronomer observing with HST to be present at either STScI or STOCC. Usually the data are organised by STScI and sent to the astronomer over the

Internet or on a computer tape. The astronomer can then start work on the images, extracting the information contained within.

Launch, first light and initial problems

In 1990, the telescope was finally ready for launch. The space shuttle *Discovery* (mission STS-31) took off with HST on board on 24 April. The HST was deployed a day later. Plate 2 shows the HST from one of *Discovery*'s flight deck windows, with the solar panels and high-gain antennae (the basic instruments of communication) all open. The HST was placed in a low Earth orbit, 600 km above the ground, where it circles the Earth every 97 minutes at a speed of 29 000 k.p.h. Astronomers waited with bated breath as systems checks showed that everything was in working order. On 20 May 1990 the moment of truth arrived and the HST turned its mirror towards the star cluster NGC 3532, 1300 light years away, for its 'first light'. As expected, the image was not quite in focus and various adjustments needed making. While the announcement was made to the world press that the HST was working, behind the scenes not everybody was happy. The images did not look right. The vast majority of the stars' light should have been focussed into a very small central point on the image known as the core, but this did not appear to be happening. The HST's advantage over ground-based astronomy was that it should provide a far smaller core than is possible from the ground. What this image seemed to show was that this core of light was surrounded by a large 'halo'. This was one worry among several for the Hubble team. The two solar panels that power the telescope were badly designed and were causing the satellite to jitter when they expanded and contracted as they heated and cooled over the course of an orbit. Also, the FGSs that should determine the position of the telescope by finding the relevant 'guide stars' were not doing their job properly. That last problem proved to have been caused by some badly written computer software, a problem that was not too difficult to solve from Earth. The other two, though, were more complicated.

The more tests that were run, the more puzzled were the astronomers. Gradually more and more of the project team became convinced that the problem lay in a serious flaw in the main mirror.

To make a telescope mirror, a circular 'blank' of very high-quality glass is meticulously ground and polished into a nearly circular shape, and then a mirror coating is applied. The specifications of this shape are vital to the focussing properties of the mirror and must be exactly right. As Isaac Newton realised when he built his first reflecting telescope, if the mirror is spherical then it will not focus the light properly. The problem is that a spherical mirror will focus light from different points on the mirror at slightly different distances from the mirror. This means that when you look at the image, it will be smeared out as not all of the light will be fully focussed at any particular point. This problem of 'spherical aberration' can be solved by grinding the mirror as a paraboloid, a not quite spherical curved surface, which is calculated to focus the light more accurately.

Unfortunately, when the HST mirror was being ground a mistake was made and the mirror was made too flat.

The main mirror was ground from a $1 million blank by the American firm of Perkins-Elmer. In order to make sure that the mirror was ground correctly the optical engineers built a device known as a 'reflective null-corrector'. This device consisted of two small mirrors and a lens hung above the HST mirror. A laser was shone on to the main mirror through the lens of the corrector, and when it was reflected by the main mirror it created a pattern of light. When the mirror was ground to exactly the right shape, this light pattern would have a particular shape that would tell the engineers that they had got it right. To work properly, the corrector's lens had to be an exact distance above the corrector's mirrors. The tiniest of errors was made, the lens being set in the wrong position by a mere 1.3 mm. The light pattern given by the corrector appeared to be inaccurate, and the technicians ground off more glass than they should have. The mirror became 2 micrometres (0.0002 cm) too flat at the edges – approximately one-fiftieth the width of a human hair!

This mistake should have been picked up during later tests of the mirror, but pressure was being put on Perkins-Elmer to deliver on time and not to exceed the budget any further. A full check therefore was not carried out before the launch, and the error was not noticed until the mirror was already in orbit. As a result the level of detail visible to the HST was reduced to a resolution of around 2 arcseconds, about the same as a good ground-based telescope. It should have been able to focus about 80% of the light from a star into the core of the image, but instead it was able to manage only 10 to 15%.

Possible solutions to the problem were considered at once. An immediate, but imperfect, solution would be to use an image-processing technique known as deconvolution. This involves working out where the light should have been and 'redrawing' the picture as it should have looked, a danger being that a lot of information contained in the original image could be lost. Every effort was then made to restore the telescope to its full potential.

The repair mission

On 2 December 1993 the space shuttle *Endeavour* (mission STS-61) took off for an eleven-day repair mission, carrying new equipment and seven astronauts who had trained for over a year. The training used the latest virtual reality technology and NASA's old standby – a full-sized mock-up of Hubble in a huge tank of water to simulate weightlessness. The first job was to rendezvous with Hubble and bring it into *Endeavour*'s cargo bay with the shuttle's recovery arm. Once this was accomplished, four of the astronauts would have to walk in space and repair and replace a number of Hubble's instruments. During the mission a record five spacewalks were undertaken, lasting for a total of just over 35 hours (a further two spacewalks had been allowed for, but the mission went so well that these were unnecessary).

The HST had been built to a modular design that would enable shuttle missions to repair or replace faulty equipment, and to swap scientific equipment to enable a wider range of observations to be undertaken. The design of the telescope included grapples by which the shuttle could capture the satellite and bring it into its cargo bay. In addition, 76 handholds were placed around the body to enable astronauts to move round easily to fix any problems. Plate 3 shows the astronaut Story Musgrave using one of these on the first of the spacewalks of the repair mission. When the HST was launched, its schedule allowed for at least three servicing missions during its anticipated 15-year lifetime.

The repair mission's main priority was to place into the telescope two special components to compensate for the faulty mirror; in addition they were to replace the solar panels and a number of broken or malfunctioning components. It would have been impossible to replace the main mirror without dismantling the telescope and then rebuilding it. As the exact nature of the fault was known, however, extra mirrors and lenses could be inserted to correct for the aberration. A new Wide Field/Planetary Camera (WF/PC2) was to be inserted in place of the old WF/PC1, and a new module to correct for the other instruments, the Corrective Optics Space Telescope Axial Replacement (COSTAR). To find room for COSTAR one of the existing instruments had to be dumped. As the High Speed Photometer (HSP) had been the most problematic, this was the instrument chosen. Plate 4 shows Kathryn Thornton lifting the phonebox-sized COSTAR out of its packaging in *Endeavour*'s cargo bay with the assistance of Thomas Akers. The actual repairs were made by two shifts of two astronauts each to reduce crew fatigue. Plate 5 shows the HST in *Endeavour*'s cargo bay towards the end of the repair mission, which was a complete success from NASA's point of view. All that remained to be seen was whether the new components would do their jobs.

At 1.00 a.m. EST on 18 December 1993 the Hubble science team gathered again to see if the new additions had solved the mirror's problems. When the image of a star appeared on the computer screens, everybody knew immediately that the repair mission had succeeded. The star's image was a bright pinpoint of light, no longer spread out by the effects of the spherical aberration. Hubble had finally fulfilled its potential.

The three pictures of the star Melnick 34, in the 30 Doradus Nebula (see plate 6), demonstrates this success. The first (6a) was taken by George Meylan of the European

PLATE 6

a   b   c

Southern Observatory, high in the Chilean Andes, at a resolution of 0.6 arcseconds, the best possible in ground-based astronomy. The next (6b) was taken by WF/PC1 before the HST repair mission, and the effects of the spherical aberration are all too clear. The central image of the star is smaller and background stars are now visible, but the defect in the mirror has caused a wispy halo to appear around Melnick 34. The vastly improved resolution of WF/PC2 makes it possible to see a wealth of faint background stars that would be impossible to detect from the surface of the Earth (6c).

The next servicing missions are planned for 1997, 1999 and 2002. The 1997 mission plans to replace the Faint Object Spectrograph with the Near Infrared Camera and Multiobject Spectrometer (NICMOS), and the Goddard High Resolution Spectrograph with the Space Telescope Imaging Spectrograph (STIS). The Advanced Camera for Surveys (ACS) is being built for the 1999 mission to replace the Faint Object Camera. These changes of instrument will vastly increase the types of astronomical observations possible for Hubble.

PLATE 3

PLATE 4

—— PLATE 7 ——

MARS

Mars, the fourth planet in the Solar System, is one of the brightest objects in the night sky. It is visibly red, even to the naked eye. This rust colour is caused by exactly that: rust. The red sands are derived from oxidised iron, and this fiery appearance led the planet to be named after the Roman god of war.

Mars orbits the Sun at a distance of about 1.5 astronomical units (about 225 million km). It is about 11% of the mass of the Earth with a radius just over half of the Earth's. Its thin atmosphere, combined with its distance from the Sun, makes the planet cold, with average temperatures near the equator of $-50°C$.

This picture was taken when Mars was close to the Earth, only 103 million km away, one of the near 'oppositions' that occurs every two years and gives us our best views of the planet. It is immediately apparent that Mars is quite cloudy. The number of clouds in the Martian atmosphere is usually quite low, and their presence indicates that the temperature of the planet has fallen recently, allowing water vapour to freeze out of the atmosphere and form clouds. The planet's north polar ice cap, which is tilted towards the Earth, is clearly visible at the top of the picture. This ice cap is made of huge deposits of water ice, very similar to the Earth's polar caps. The southern polar cap on Mars, however, is very much colder as it is pointed away from the Sun. This cap is mostly frozen carbon dioxide. The red spot on the left-hand edge of the picture is the huge volcano Ascraeus Mons, protruding through the clouds. Beneath this is a darker patch, the Valles Marineris, a huge rift valley system that is 5000 km long and up to 500 km wide in places. It is thought that this huge scar on the surface of Mars may be connected to the mysterious upheaval that created the Tharsis Bulge, an area of ground that has been pushed as far as 10 km above the surrounding crust.

The picture was taken on 25 February 1995 with the small, high-resolution planetary camera section of WF/PC2.

—— PLATE 8 ——

JUPITER

Jupiter is the fifth and largest planet of the Solar System. It is over 300 times more massive than the Earth and has over 1300 times the Earth's volume. The mass of Jupiter is more than twice that of all the other planets combined; no wonder it was named after the king of the gods. Jupiter orbits the Sun at a distance of about 5.2 AU, completing an orbit every 11.9 Earth years. Surprisingly, despite its size, Jupiter has the shortest 'day' of any planet, just under ten hours.

Considering the size of Jupiter, its mass is actually quite low. The average density of Jupiter is only 1.3 times that of water (compared to Earth, whose density is about 3 times that of water). The reason for this is that Jupiter is composed mainly of gas. Together with Saturn, Uranus and Neptune it is one of the 'gas giants'. Like the Sun, most of Jupiter is hydrogen and helium, both very light elements, hence its low density.

Jupiter's surface appears to have alternate light and dark bands of colour parallel to its equator. These are known as 'zones' and 'belts'. Observations in the infrared show that the zones are cooler than the belts, which means they must be higher up in the atmosphere. The zones are points where high-pressure regions in the lower atmosphere rise up, and the belts are low-pressure regions where the gas falls back. This is a convection effect, taking heat from the planet's interior and passing it out into space, the banded structure being a result of Jupiter's very fast rotation. The bands are different colours due to the different elements and molecules present in them and their different temperatures.

This rapid rotation also encourages massive storms to form in the upper atmosphere. The most famous of these is the Great Red Spot, just south of the Jovian equator. The Great Red Spot was first seen in 1831 and has persisted ever since. It measures 15 000 × 20 000 km – large enough to swallow the Earth ten times. Observations show that it is slightly cooler than, and about 8 km higher than, the surrounding clouds.

This picture was taken by WF/PC2 on 13 February 1995, when Jupiter was 961 million km away. The image clearly shows the Great Red Spot on the right, with three white storms to its south-west. These storms have been observed for about 50 years, during which time they have evolved considerably. They have been getting closer together while also moving further west from the Great Red Spot. The white colour of the clouds is caused by ammonia high in the atmosphere, which has been dragged from lower in the atmosphere by the storms. As the ammonia gets higher it cools until eventually it freezes and forms white ice particles.

—— PLATE 9 ——

GALILEAN MOONS

Jupiter's four largest moons are named collectively after Galileo, who discovered them in January 1610 (although they were also independently observed by Simon Marius at about the same time), their individual names being derived from figures in Roman mythology. All four can be seen with a small telescope or even a pair of binoculars, and the innermost of them can even be seen to move over the course of a few hours.

The innermost of the four is Io. It orbits once every 1.77 days. Slightly larger than our Moon, Io is 442 000 km from Jupiter, and huge tidal forces warp and heat the entire moon. Io is the only other body in the Solar System known to have active volcanoes (those on other planets are all thought to be extinct), and their outgassings give Io a thin atmosphere.

The next moon out is Europa, the smallest of the four, at about two-thirds the mass of our Moon. Europa's surface consists mainly of frozen water ice criss-crossed by cracks and fissures. Europa is virtually free from impact craters, which implies that its surface is young, having been active since the era of major bombardment by asteroids and meteorites came to a close. The HST has found a thin atmosphere of oxygen around Europa.

Ganymede is the third Galilean moon and the largest in the Solar System (just over twice as heavy as the Moon) and is virtually a planet in its own right. It orbits Jupiter every 7.2 days at a distance of over 1 million km. The surface is cratered and rather flat. There are also large faults and cracks on the surface, probably as a result of the tidal force of Jupiter causing stresses in the moon. The HST has been able to find ozone (a molecule made up of three atoms of oxygen joined together) on the surface of Ganymede.

Callisto is the outermost Galilean moon, nearly one-and-a-half times the size of the Moon, which it closely resembles in being heavily cratered, with bright and dark patches. The surface is probably made up of a mixture of rock and ice, and in ultraviolet light the HST has seen fresh ice on Callisto. This ice has formed when tiny meteorites or energetic particles from Jupiter collide with the moon and melt a small amount of the dirty ice on the surface, which then refreezes.

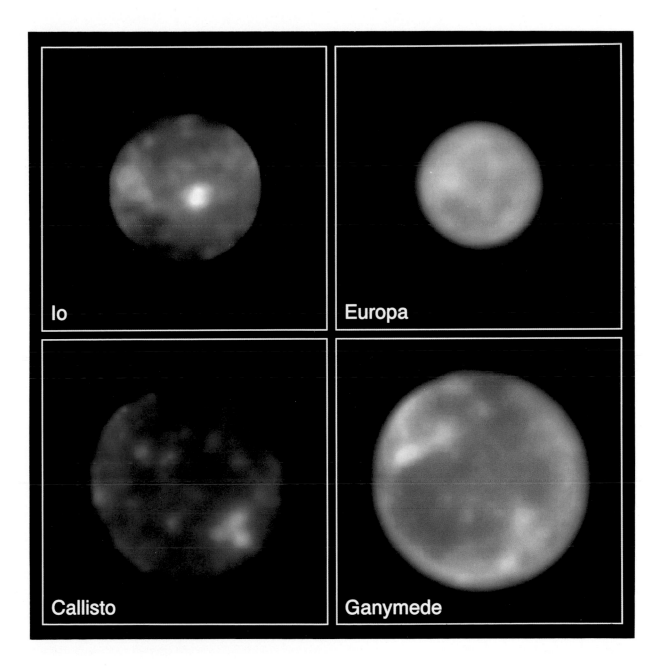

—— PLATE 10 ——

SATURN

Saturn is the second largest planet in the Solar System, named after the Roman god of agriculture. It orbits the Sun every 29.5 Earth years at an average distance of 9.5 AU. Its best-known feature is its series of rings, discovered by Galileo in 1610, which are made up of small icy chunks (mostly just a few centimetres across). One of the 'gas giants', it is primarily composed of hydrogen and helium.

Aurorae occur on all planets with a magnetic field, including the Earth. Charged particles such as electrons and protons from the highly energetic solar wind are channelled downwards by the planet's magnetic poles, and collide with gas molecules in the atmosphere, causing them to emit radiation and giving rise to spectacular luminous curtains. On Earth, these are occasionally visible, even from middle latitudes: the Aurora Borealis around the north pole and the Aurora Australis in the south. In the Saturnian atmosphere, however, most of the gases (mainly hydrogen) radiate in the ultraviolet region of the spectrum, invisible to ground-based instruments because most ultraviolet radiation does not penetrate the Earth's atmosphere. The HST's ultraviolet capability enabled pictures to be taken of the aurorae at Saturn's north and south poles.

The top picture, taken in ultraviolet light, clearly shows the aurora at Saturn's north pole. The 'auroral curtain' of light pushes as high as 2000 km above the tops of the Saturnian clouds. Even though Saturn's south pole is hidden from view, the south auroral display is just visible around the edge of the planet. The curtain varied rapidly during the time of the observations, but the brightest point always stayed at a set angle from the Sun.

The bottom picture shows Saturn in visible light, and the differences are striking. Ultraviolet sunlight is reflected from higher in the Saturnian atmosphere than visible light. The parts from which ultraviolet light is reflected do not contain the banding and structure present in the lower atmosphere, giving Saturn a featureless appearance in the higher frequency ultraviolet. Also present in visible light is a huge white storm system near the equator which the HST was able to track while it raged.

—— PLATE 11 ——

TITAN

Titan is the second largest moon in the Solar System. Discovered in 1655, it orbits Saturn every 16 days. At 5150 km in diameter, Titan is larger than Mercury. It has a thick atmosphere, primarily nitrogen (99%), the rest being methane and other hydrocarbons. Titan's surface pressure is one-and-a-half times higher than that of the Earth, and the surface temperature is around $-180°C$. The atmosphere shrouds the moon completely in a featureless orange haze, similar in composition to the poisonous smog that cars and factories produce on Earth.

Titan is the only other body in the Solar System that may have oceans and rain. Due to the very low temperature of this moon, its oceans are likely to be of an ethane–methane mixture rather than of water. Titan may well be similar in its conditions to Earth before life began.

This infrared image was taken by WF/PC2 in October 1994. Looking in the infrared the HST was able to penetrate the atmosphere and view the surface for the first time. The resolution was good enough to allow features as small as 570 km across to be seen. The bright patch on the left of the picture is about the same size as Australia, and may be a continent or a sea. The HST's examination also confirmed that Titan is tidally locked to Saturn, the same face always pointing towards the giant planet. The picture shows the hemisphere of Titan that always leads in its orbit around Saturn. This is also the brightest (most reflective) face of the moon.

The HST's images of Titan will provide important information in the planning of the *Cassini* space probe, which is intended to put the Huygens lander on the surface of Titan early next century.

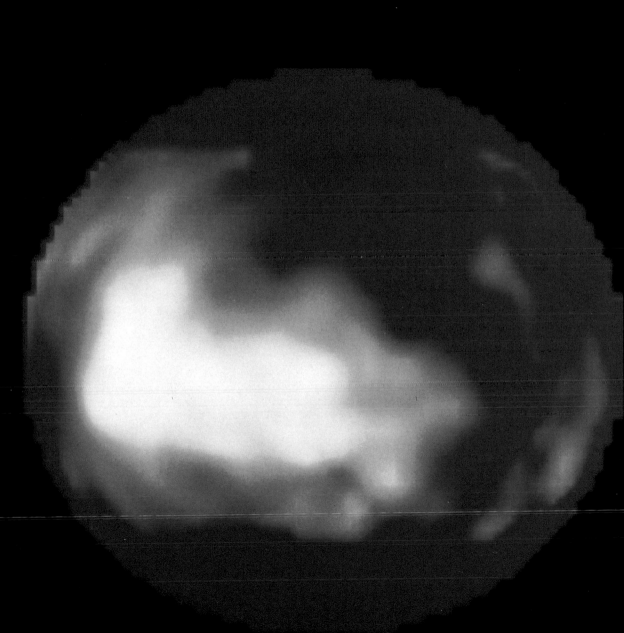

—— PLATE 12 ——

PLUTO AND CHARON

The most remote planet in the Solar System is Pluto, which has a moon, Charon. Pluto was discovered in 1930 by Clyde W. Tombaugh and was named after the god of the underworld because of its distance from the Sun. Pluto's highly eccentric orbit takes it between 30 and 49 AU from the Sun; for most of the time in its 249-year orbit it is the most distant planet from the Sun (although it sometimes comes inside the orbit of Neptune). The existence of Charon as a separate entity was established by James Christy in 1978. Once the identity of Charon was confirmed, the mass of the pair could finally be calculated accurately: Pluto was found to be about one five-hundredth of the mass of the Earth, and Charon is about ten times smaller than this. The pair are only 17 500 km apart on average and each always has the same face pointing to the other, swinging around once every 6.4 days. Due to their enormous distance from Earth, they are virtually impossible to resolve as separate objects with a ground-based telescope.

Using the Faint Object Camera, the HST took this image on 21 February 1995 when Pluto was only 4.4 billion km away (about 30 AU), nearly as close as it ever gets. The image is in 'false colour' so that certain details are enhanced. It is the first clear picture to show Pluto and Charon as separate, clearly defined discs (the images available to Christy gave Charon the appearance of a lump on Pluto's side). From this picture the diameters of the two could be accurately measured for the first time, Pluto at 2320 km and Charon at 1270 km. The images show a small amount of detail on the surfaces, including a bright band above the equator of Pluto. Pluto appears very bright for its size and probably has a reflective, icy surface, possibly methane ice. Charon is redder than Pluto, suggesting that it has a different surface composition.

—— PLATE 13 ——

COMET HALE-BOPP

Comet Hale-Bopp was first sighted beyond the orbit of Jupiter by two amateur astronomers, Alan Hale and Thomas Bopp, on 23 July 1995. Its discovery has excited many, both inside and outside the astronomical community, as it is so bright now that it could well be the most spectacular comet of recent times when it comes closer to the Sun.

For most of the time comets are frozen solid, but when they pass close to the Sun they heat up and the frozen gas will begin to boil away. It is when this starts to happen that comets become visible from Earth. As the comet is heated by the Sun a stream of material spews out to form a tail. This tail always points away from the Sun as it is being driven away from the comet by the solar wind (a constant stream of fast-moving particles ejected by the Sun). Each time a comet circles the Sun it loses some of its mass and becomes less bright.

This image of Hale-Bopp was taken while it was still further away than Jupiter (around 1 billion km away), using WF/PC2. The lower picture shows stars in the background in an elongated form because the HST was tracking the movement of Hale-Bopp, which changed position slightly during the time of the exposure. The upper picture has been processed by computer to enhance the image of the comet. The stars have been removed and the image magnified so that the individual pixels of the WF/PC's CCD image are visible. At the distance of the comet the HST was able to resolve details as small as 500 km across. The enhanced image clearly shows a large blob of material being ejected by the nucleus of the comet, at a little over 100 k.p.h. The blob is spinning round and away from the comet at the same time, creating the spiral pattern. If the comet is made up of ices that boil quickly and violently, that would explain why it is so bright despite being so far from the Sun. Unfortunately, if this is the case then Hale-Bopp may well have fallen apart or even boiled away completely before it gets close enough to the Earth for us to see it with the naked eye.

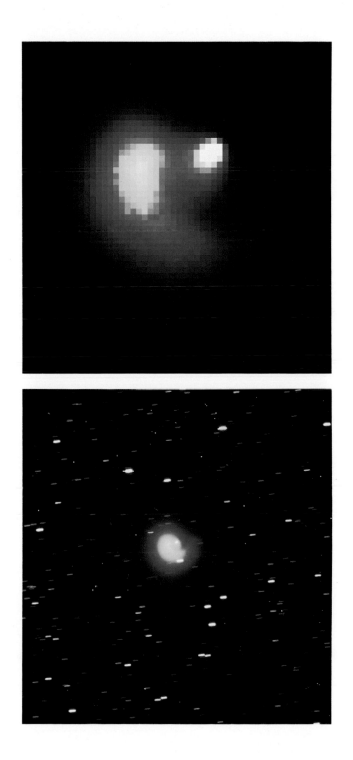

—— PLATE 14 ——

COMET P/SHOEMAKER-LEVY 9

Comet P/Shoemaker-Levy 9, the ninth comet to be discovered by the team of Carolyn and Gene Shoemaker and David Levy, in March 1993, would probably have passed unnoticed had it not come close to Jupiter. Jupiter's tremendous gravitational pull caused a tidal force on the comet which tore it apart. This red light image from WF/PC2 shows 21 fragments extending over 1.1 million km of space, three times the distance from the Earth to the Moon. The largest of these fragments were estimated to be 2 to 4 km across.

The fragmentation occurred about eight months before the comet's discovery. When the orbit of the comet was calculated it was realised that it would impact with Jupiter in mid-July 1994 (see plate 15). The interaction with Jupiter that had ripped the comet apart had also captured it in Jupiter's intense gravitational field, and the comet orbited the giant planet a few times before crashing into it.

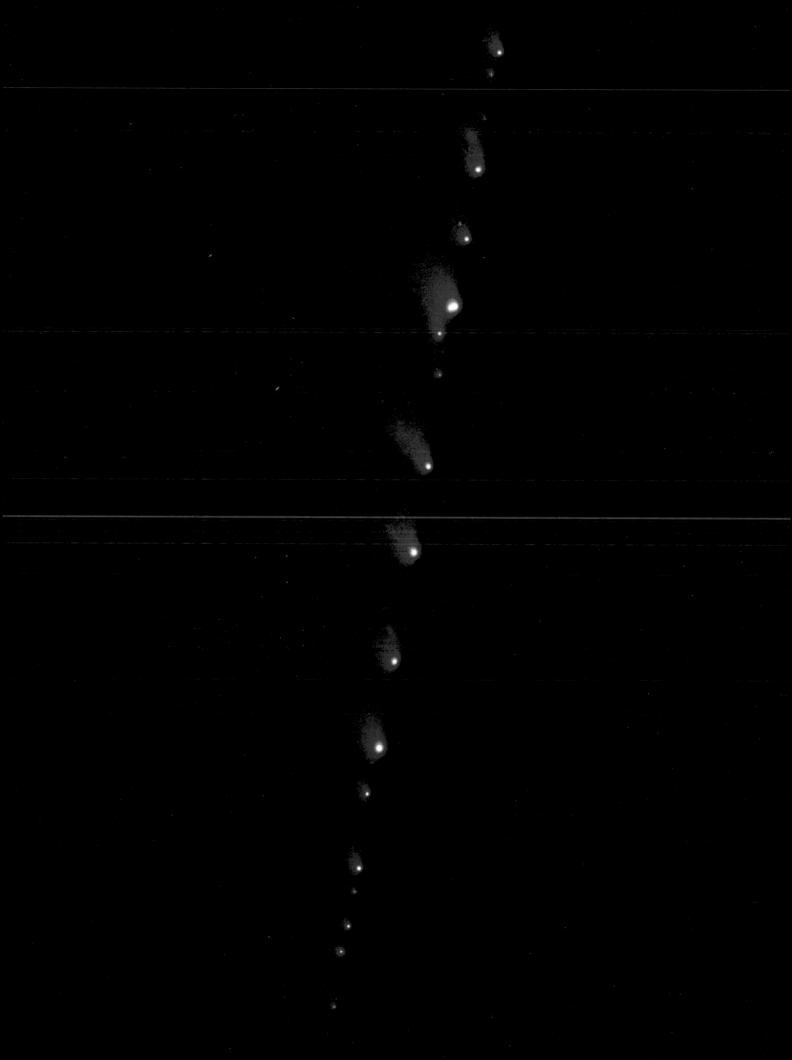

—— PLATE 15 ——

IMPACT SITE OF FRAGMENT G OF COMET P/SHOEMAKER-LEVY 9 ON JUPITER

On 18 July 1994 fragment G of comet P/Shoemaker-Levy 9 entered the Jovian atmosphere and began to burn up. Fragment G was the brightest of the comet's 21 fragments (see plate 14) and so probably the largest. The scar it left on the planet was about the size of the Earth and was caused by an explosion equivalent to 10 million megatons of TNT. Over the five days that comet fragments were colliding with Jupiter the planet received an enormous pummelling of 100 million megatons of TNT (equivalent to more than 10 000 times the destructive power contained in the world's nuclear arsenal at the height of the Cold War).

Unfortunately, the impacts occurred on a region of Jupiter just out of view from the Earth at that time. Telescopes had to wait approximately four minutes for Jupiter's rotation to bring the impacts into sight. This image was taken an hour and three-quarters after the impact, by which time the scar had travelled quite some way around the planet. The large blemish is the G impact site, and slightly above it is the remnant of the impact of fragment D several hours earlier.

The black scar on the clouds is the result of the explosion of fragment G. The dark material is probably fine dust which condensed after the explosion of the comet fragment, either from the comet itself or from the Jovian atmosphere. The further curve of material below the circular scar was formed as fragment G entered the atmosphere at an angle of about 45° and was formed from debris from the fireball resettling in the atmosphere. The shock wave caused by the explosion was then still travelling at 1800 k.p.h.

—— PLATE 16 ——

ORION NEBULA

The Orion Nebula is in the constellation of Orion, the Hunter, one of the best-known constellations in the sky and one of the few that looks at all like its namesake. The fuzzy patch below the three stars that form Orion's Belt is the Orion Nebula itself (Orion's Sword). Located a mere 1500 light years away, the Orion Nebula is one of the closest, and so one of the most spectacular, star-forming regions. This picture shows at least 700 stars in the early stages of their lives.

Nebulae often contain enough gas to make millions of stars the size of our Sun, although not all of this gas will develop into stars. At first what will become stars are just parts of the cloud that are slightly denser than the surrounding regions. As they are denser, they attract more gas from around them by their gravitational pull. As these 'proto-stars' grow the temperatures and pressures at their centres also increase. Once the temperature at the centre reaches 10 million °C or more it is hot enough for nuclear fusion to begin. Before the centre becomes this hot, two hydrogen atoms will bounce off each other every time they collide, but at this temperature or above some of them fuse together, giving off energy as they do so. This energy comes from Einstein's famous $E = mc^2$ equation, as when the atoms fuse together some of their mass is lost and becomes energy. In the centre of young stars four hydrogen atoms join together and form a helium atom. The energy created by this process is released as light which works its way out of the star, causing the star to shine. Any proto-star that grows to one-tenth of the size of the Sun or more will be able to get hot enough to start nuclear fusion.

Some nebulae have no star formation inside them but some, such as the Orion Nebula, are filled with many young stars. This star formation occurred in the Orion Nebula about 300 000 years ago, a very recent event in astronomical terms. The nebula is lit up by the stars inside it. As they shine, the stars put a lot of light into space. Some of this light hits atoms in the gas of the nebula. If this gas is in the right condition, it can absorb and then re-emit some of this light. The frequency of the re-emitted light then tells us what kinds of atom have emitted it.

This picture, taken with WF/PC2 over the course of 15 months with three filters in visible light, is a composite of 45 separate images. The colour of the gas shows what kinds of atom were emitting in that area: hydrogen is green, oxygen is blue and nitrogen is red. Very large, hot stars give out so much light that they push the surrounding gas away from them, 'blowing away' a part of the nebula. This is what has happened in the Orion Nebula, and we are fortunate in having a viewpoint from which we can look down into the heart of the nebula to the four stars known as the Trapezium, which have pushed out gas. These stars are the most massive in the nebula, having been formed from the large concentrations of gas at the centre.

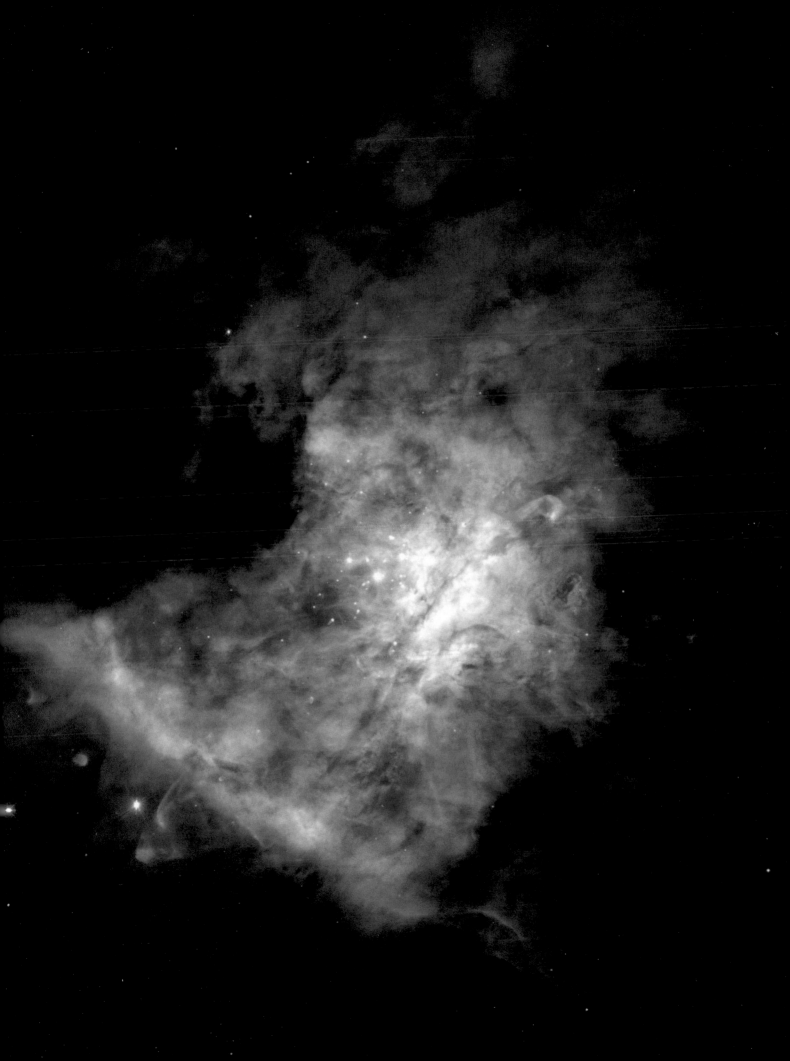

—— PLATE 17 ——

EAGLE NEBULA (M16)

The Eagle Nebula is another star-forming nebula much like the Orion Nebula. At 7000 light years away in the constellation of Serpens, it is much further away than the Orion Nebula. Its alternative name, M16, is its designation in Charles Messier's catalogue of 1781. Messier compiled a list of objects in the sky that were 'fuzzy' and extended, unlike the pin-point images of stars. Objects with an M number (which are nebulae, star clusters or galaxies) can be seen with a small telescope or binoculars, or even the naked eye.

These pictures, taken with WF/PC2 on 1 April 1995, show vast columns of cool, dense gas about a light year in height. These columns are so dense that the stars within them are hidden from sight, their light unable to penetrate the gas surrounding them. The columns are mostly molecular hydrogen, two hydrogen atoms joined together, which is normally too fragile to exist outside of nebulae. This fragility arises from the fact that energetic light (such as ultraviolet light) is able to split the molecules when it hits them. The columns also contain a lot of microscopic particles mostly made of carbon, known as 'dust', that are able to form in the protective environment of a nebula.

The columns have such a strange shape because there are a few very young, massive stars just off the top of the picture which are 'blowing away' the gas and dust (in the same way as the Trapezium stars in the Orion Nebula). Their light is breaking up the molecular hydrogen and heating up the gas in the cloud. When the gas gets hotter, it starts moving faster and eventually escapes from the nebula. The less dense parts of the nebula are evaporated by the stars' radiation first, leaving behind these columns.

In this image, again, the colours of the gas indicate which atoms are emitting light in that area. This time red is sulphur, green is hydrogen and blue is oxygen.

Eagle Nebula (M16)

—— PLATE 18 ——

EAGLE NEBULA (DETAIL)

This picture shows a detail from the top of the largest column of gas on plate 17. The high resolution image shows small elephant trunk-shaped 'blobs' of gas, about the diameter of the Solar System, remaining at the tips of the columns after the 'stellar winds' of the larger stars have blown away much of the gas. These blobs have been christened EGGs (for evaporating gaseous globules).

These EGGs are thought to be the birthplaces of stars in the nebula. They are small, dense concentrations of gas that produce stars inside nebulae. They are so much denser than the surrounding gas in the nebula that the massive stars which are eroding the nebula have not yet managed to blow them away. It is thought that some, maybe many, of these EGGs have stars inside them which will slowly be revealed as the EGG evaporates. Some of these stars might not have reached the point where nuclear fusion can begin; others may contain the beginnings of planetary systems.

EAGLE NEBULA (DETAIL)

———— PLATE 19 ————

TARANTULA NEBULA WITH STAR CLUSTER R136

This image shows the Tarantula Nebula (otherwise known as the 30 Doradus Nebula), and at its heart the star cluster R136. They are both located in the Large Magellanic Cloud (LMC). The LMC is a small companion galaxy to the Milky Way that can be seen from the Southern Hemisphere. The LMC has a very irregular shape as the Milky Way is slowly pulling it apart, and eventually it will fall into our Galaxy. It was called the Magellanic Cloud (together with its near neighbour, the Small Magellanic Cloud) after the explorer Ferdinand Magellan who was the first person from Europe to report seeing the two galaxies, in 1519. The LMC is very, very close in intergalactic terms at around 160 000 light years away.

These pictures were some of the first to be taken using WF/PC2 after the repair mission, and they show detail as small as 25 light days (a light day being about 26 billion km) across.

The Tarantula Nebula is an H II region. H II regions are clouds of gas made up of ionised hydrogen. Normally hydrogen atoms have one proton and one electron orbiting this proton. When hydrogen is ionised, ultraviolet light causes the electron to be lost from the proton. The star cluster at the centre of the nebula is giving out so much ultraviolet light that this has happened to almost all of the hydrogen atoms in the cloud. The protons and electrons in the cloud are very energetic, and cause the nebula to glow very brightly.

TARANTULA NEBULA WITH STAR CLUSTER R136

—— PLATE 20 ——

TARANTULA NEBULA (DETAIL)

The Tarantula Nebula is of particular interest because of the stars that have caused the H II region to form. In most star-forming regions (such as the Orion or Eagle nebulae) only a few per cent of the gas turns into stars, and most of these stars are small stars, the size of the Sun or less. In the Tarantula, however, much more gas turned into stars and many of these stars were very massive, several times the size of the Sun. The larger a star is, the hotter it is, so these stars have been able to ionise a huge portion of the Tarantula Nebula.

The star cluster at the heart of the nebula is known as R136. Astronomically, it is very young, only a few million years old. Because it is so young none of the stars in R136 have had time to evolve. It still contains many huge stars, each producing millions of times more energy than the Sun.

Because there are so many huge stars in a very small region of space, it was originally thought that R136 was one supermassive star several hundred times the size of the Sun. It was possible from the ground to resolve R136 into a few objects, but each of these would still have had to be more massive than any other star of which we know. With the HST's improved resolution it was possible to show that R136 is, in fact, a cluster of at least 3000 separate stars. It is also possible to look at these stars and analyse them individually.

—— PLATE 21 ——

STAR CLUSTER NGC 1850

NGC 1850 is a star cluster about 166 000 light years away and, like 30 Doradus, it is located in the Large Magellanic Cloud. This picture was taken with WF/PC2 and shows a region of the LMC 130 light years across. The picture is a composite of filtered images taken across the whole range of the HST's sensitivity from infrared to ultraviolet. The HST has been able to make out around 10 000 separate stars in this picture and provide information about all of them.

The HST has found that the stars in this cluster are of three different ages. How long a star will live depends upon how massive it is. Large stars burn their nuclear fuel faster than smaller stars. This means that massive stars are hotter and burn brighter, but do not live as long as smaller ones. A star ten times larger than the Sun will live for only about 10 million years, while a star the size of the Sun would remain much the same for 10 billion years or more. The hotter a star is, the bluer it looks. Combining these two properties of stars means that the age of a group of stars can be estimated by seeing how hot (blue) the brightest (largest) stars are. The larger a star that is still shining, the younger the cluster.

About 20% of the stars in the picture are young, massive, white stars which have surface temperatures of up to 25 000°C. Some of these stars are so large that they can only be about 4 million years old. Around 60% of the stars belong to the large cluster NGC 1850 itself. These are yellow and are probably about 50 million years old. The stars in NGC 1850 are around 200 light years in front of the very young stars. The rest of the stars in the picture are old, red stars that belong to the background of stars in the LMC.

Like R136 (see plates 19 and 20), NGC 1850 is a cluster of stars that all formed at almost the same time in a large nebula. NGC 1850, however, has pushed all of the remaining gas of the nebula away. The stars have managed this by a combination of their light and some of the very massive stars exploding in supernovae which removed a lot of gas from the cluster. It is thought that the gas was expelled at high speed, and that when the expanding gas shell collided with other gas clouds 200 light years away it initiated a burst of star formation in the other clouds by compressing them. This has caused the spread of ages and colours of stars in the picture.

—— PLATE 22 ——

STAR CLUSTER M15

M15 is one of about 150 large star clusters that orbit the Milky Way. On account of their spherical appearance, they are known as globular clusters. M15 is 32 000 light years away and orbits the centre of the Milky Way at a distance of 30 000 light years every 300 million years. It contains several million stars, most of which lie within 40 light years of the centre of the cluster, which has a total mass equal to nearly a million Suns.

Globular clusters are estimated to be between 10 and 20 billion years old, which makes them among the oldest objects in the Milky Way. They are thought to have formed from huge clouds of gas that were falling into the very young Milky Way in the same way as stars in the Tarantula Nebula (plates 19 and 20) and NGC 1850 (plate 21) formed. Possibly as much as half the gas in the clouds turned into stars, compared with just a few per cent in such star-formation regions as the Orion Nebula. It is only because so many stars formed in such a small region that they still exist today and have not been pulled apart by the tidal pull of the Milky Way.

The picture, taken by WF/PC2 in 1995, shows a region 28 light years across. The enlarged detail shows the core of M15, only 1.6 light years across. This picture of the core is a composite of visual and ultraviolet light images in which the colours of the stars correspond to their surface temperatures, the blue end of the spectrum indicating the hotter stars.

M15 is one of around 20 globular clusters that have a peculiar distribution of stars, those in the centre being far more concentrated than astronomers would expect. By measuring the speed at which stars near M15's centre are moving, the HST was looking to see whether this was caused by a large black hole at the centre or by a strange phenomenon known as 'core collapse'. As the HST did not find a black hole, the concentration must be due to core collapse. The gravity of the stars in M15's core attracts the stars, which then fall into the core. Given certain conditions, the centre can become so dense that the stars begin to form binary pairs; as they cannot get any closer to each other, the core collapse is eventually stopped.

—— PLATE 23 ——

JETS FROM YOUNG STARS

Young stars often produce jets of material streaming out from their poles. These jets, Herbig-Haro 1 and 2, provide a number of clues as to what is happening during star formation. In this picture of the jets, the star itself is invisible, hidden inside a thick cloud of dust and gas from which it is forming, while the jets stretch half a light year in either direction. Stars with jets like this are all known as Herbig-Haro objects after the astronomers who first investigated the strange jets of glowing gas with no apparent source. This young star is in the region of the Orion Nebula, about 1500 light years away, but these objects can be found almost anywhere where stars are forming.

As gas falls inwards to form a star it is rotating slowly. In the same way that ice skaters execute spins by starting slowly, and increasing their speed of rotation through bringing their arms in, so the rotation of the infalling gas quickens as the gas gets closer to the star. This gives the gas a lot of kinetic energy. The gas starts to collide with other infalling gas, and in doing so loses some of this energy and falls into an 'accretion disc'. In this disc it continues to lose its energy by friction with the other gas in the disc; the lost energy is then radiated away as light. As it loses its energy, and hence speed, it falls further inwards until finally it falls onto the star itself in the centre of the disc. Some (possibly most) of the material falling onto the star is heated up again and thrown out along the axis of rotation of the star. Often these ejections of material are sporadic, making the jet look like a lumpy necklace. Quite why this ejection happens, and especially why the jets are so thin, is unknown. It may be that the magnetic field of the star has something to do with this, and the jets flow out along the lines of magnetic force from the star's magnetic poles. These jets remain for the first hundred thousand years or so of the star's life before the star finally blows away its disc and material stops falling onto it.

These jets then flow out into space at speeds of up to 1 million k.p.h. Eventually the jets hit cooler, denser gas in the interstellar medium. This impact forces the jets to slow down immediately, which causes the gas to heat up as the vast energy of the jet due to its speed is lost. As the newly heated gas cools down again it loses its energy as light, the colour of this light depending on what atoms are present in the gas. At the end of the jets are 'bow shocks', very similar to those produced by a boat speeding through water.

These pictures were taken by WF/PC2 in 1995.

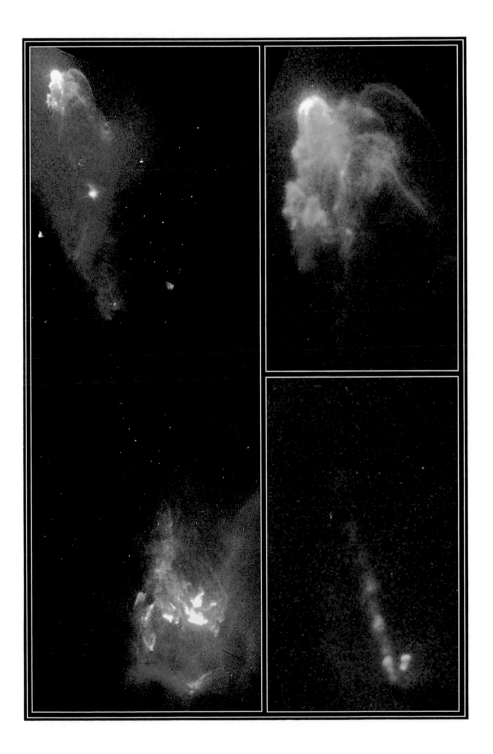

—— PLATE 24 ——

PROTO-PLANETARY DISCS IN THE ORION NEBULA

Proto-planetary discs, or 'proplyds', are thought to be the basis for what will eventually become a planetary system much like our own Solar System. In the Orion Nebula as a whole 153 proto-planetary discs have been observed. This implies that the formation of planetary systems may not be an uncommon occurrence. The discs shown here are silhouetted sharply against a very bright region of the nebula. The cool, red, central star of the disc is very obvious. These central stars are normally about the size of the Sun. The sizes of the discs range from two to seven times the size of the Solar System.

The planets of our Solar System are thought to have formed from a similar disc of gas and dust that surrounded the Sun just after its formation. Within this disc small clumps of matter, or planetesimals, formed and grew under the action of their own gravitational attraction. Some of these planetesimals would then be attracted together to form the building blocks of the planets. Eventually, as more and more came together, objects the size of planets would evolve. In the inner parts of the Solar System the radiation from the Sun would blow away most of the light elements such as hydrogen and helium, leaving planets formed mainly of carbon, oxygen and metals such as iron. These planets became the rocky planets such as Earth. Further out the Sun would not be powerful enough to blow away the light elements before the gas giants like Jupiter had formed. The remains of the planetesimals that did not aggregate into planets are found in the Solar System as asteroids and comets. It is the early stages of this process that it is believed is being observed around these stars.

This theory of the formation of the Solar System has been postulated by planetary scientists and astronomers for many years. It is not until now, with the power of the HST, that possible examples of proto-planetary discs have been seen. These pictures are close-up details from the WF/PC2 images of the Orion Nebula. The regions in these images are only 30 times the size of the Solar System across and show young stars very soon after their birth. What is of great interest is that these stars show large discs of gas and dust around them (99% gas and 1% dust).

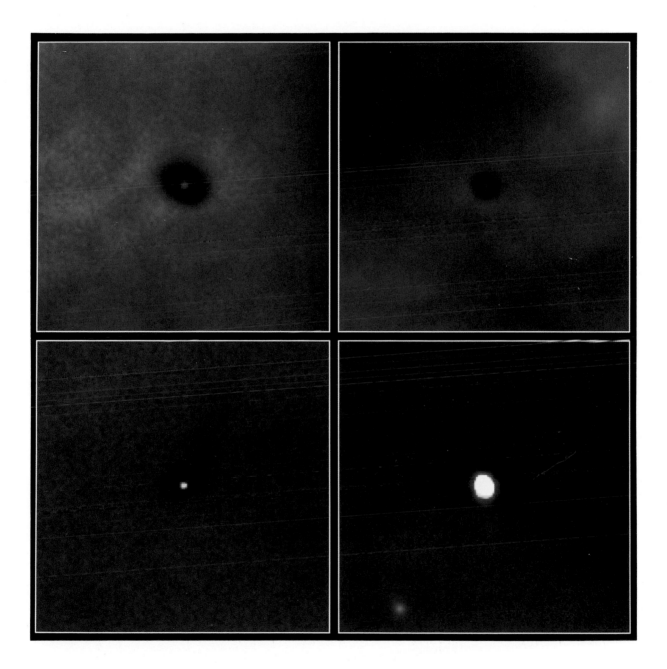

—— PLATE 25 ——

GLIESE 229 WITH BROWN DWARF COMPANION

The star Gliese 229 (or GL 229), 18 light years away in the constellation Lepus, is a small, cool star of a type known as a red dwarf. To the right is a tiny companion called GL 229B, which cannot really be called a star because it is not massive enough to be able to burn hydrogen fuel. GL 229B is the first to be definitely identified as belonging to a class of objects known as 'brown dwarfs'. Brown dwarfs fill the gap between planets and stars. They are far larger than planets but not large enough to begin nuclear fusion in their cores and become stars. Some astronomers think that the Milky Way is full of billions or trillions of brown dwarfs. Because they do not burn hydrogen, however, they are very difficult to see as they do not shine anything like as brightly as stars do. When a gas cloud contracts it heats up. If it is too small to form a star and becomes a brown dwarf it will still be hot. However, its surface temperature will be at most hundreds of degrees (the surface of some stars may be heated to thousands or even tens of thousands of degrees). That heat will be radiated slowly into space, but as low energy infrared radiation. The gas giants of our Solar System, especially Jupiter, do the same, but their heat radiation is tiny compared with the Sun. It is this infrared light coming from GL 229B that the HST has been able to see.

GL 229B is 20 to 50 times the mass of Jupiter, but it is far denser, having only the same diameter. It is at least 6.5 billion km from GL 229, further than Pluto is from the Sun, a vast distance enabling the HST to resolve GL 229B as a separate object. Despite being masked, the light from GL 229 nevertheless fills much of the picture. GL 229B is the faintest object ever to be seen around a star other than the Sun. Its existence adds more weight to the idea that planetary systems can form around other stars, although any object smaller or closer to a star than GL 229B would be too faint to see. The line across the picture from the star is not real. It is an artefact of the imaging of any bright object and is known as a 'diffraction spike'.

——— PLATE 26 ———

ETA CARINAE

Ten thousand light years away in the Keyhole Nebula in the constellation Carina is one of the most massive and unstable stars that is known. Eta Carinae can be one of the brightest stars in the sky, 4 million times brighter than the Sun and approximately 150 times larger; its surface temperature is probably over 29 000°C (five times hotter than the surface of the Sun). Eta Carinae is known to have large variations in its brightness over very irregular periods. Its most extreme variation occurred between 1835 and 1845, when it suddenly brightened and became the second brightest star in the sky.

These variations in the light from Eta Carinae are caused by the extremely unstable nature of the star. It is constantly ejecting material, and sometimes this ejection is very sudden and massive. Its instability is derived from its size. A star as huge as Eta Carinae uses up its supplies of nuclear fuel rapidly and will probably not be able to survive for more than a few million years. While it is burning fuel, however, it is producing so much energy that it cannot all be radiated as light; some of the energy is therefore lost in the ejection of matter.

The colours in this picture are the actual colours of the light that HST detected in Eta Carinae. The image was taken using WF/PC2. Previous images of Eta Carinae had been taken with WF/PC1 before the servicing mission, but not at a resolution that showed the detail seen in this picture.

This picture shows the shell of the material that Eta Carinae ejected in the outburst of 1835–45. This shell is the outermost red, glowing ring about the star. Some of this gas is still travelling at over 3 million k.p.h. The star brightened when it ejected the material because the ejecta was still very hot and so radiated light. As the shell expanded it cooled by radiating light, which was observed as the star getting brighter. This shell is rich in nitrogen, oxygen, carbon and other elements that are formed by nuclear fusion in very massive stars.

Eta Carinae has now got dimmer again, probably due to a shell of dust blocking some of the light from the star. This shell is the innermost blue-white region, named the Homunculus Nebula because of its rather grotesque shape. This region has two lobes of material travelling in opposite directions. Unlike young stellar jets from Herbig-Haro objects (see plate 23), these ejections appear to be in the plane of the disc, quite the opposite of what would be expected.

—— PLATE 27 ——

NGC 7027 NEBULA

The planetary nebula NGC 7027, 3000 light years away in Cygnus, is the result of the ejection of matter from a red giant or supergiant star. Planetary nebulae were so named in the eighteenth century, before their true nature was established, for when observed through telescopes of the time they showed a disc similar in appearance to a planet. They are normally very young in astronomical terms, mostly less than 50 000 years old.

This HST picture is a composite of images in visible and infrared light. The results of the different stages of mass expulsion from the star are clearly visible: the faint blue concentric rings are the shells of gas that were gradually ejected by the star while it was still burning helium. Once the core had exhausted its helium, it suddenly ejected all of the remaining matter around the core in the bright burst visible at the centre of the picture.

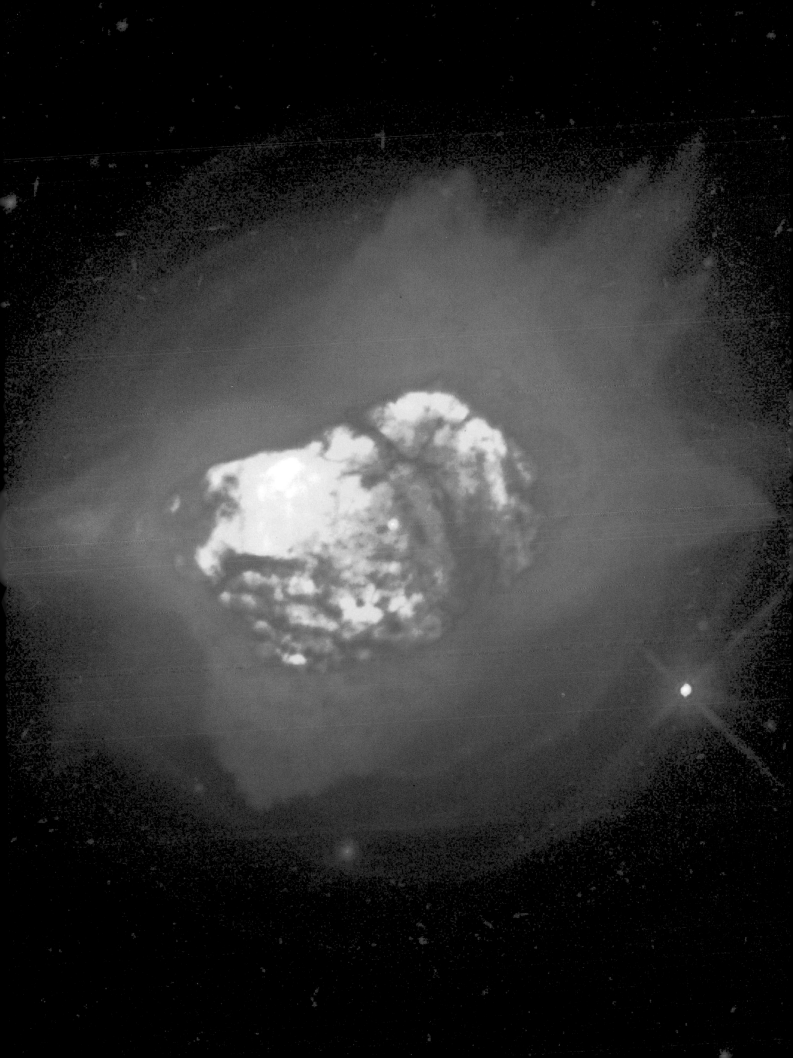

—— PLATE 28 ——

EGG NEBULA

The Egg Nebula (CRL 2688) is a planetary nebula 3000 light years away. Its central star is a red giant which is currently ejecting gas. The final ejection phase of a red giant lasts 1000 or 2000 years, a period equivalent to five or ten minutes in the life of a human being.

This picture was taken in red light by WF/PC2 in 1995. The gas is being thrown out from the central star at around 190 000 k.p.h. The star itself is hidden behind a dense band of dust and gas which can be seen crossing the centre of the nebula, presumably in the plane of the star. Gas is being ejected from the star in bursts; these occur between 100 and 500 years apart and can be seen as arcs in the smooth background of the nebula. The appearance of the ejecta as arcs rather than circular shells is probably due to dense blobs of material in the nebula that cast shadows upon some parts of the shells. This is especially obvious in the dark plane of the star, giving the nebula the hourglass shape characteristic of many planetary nebulae. The four striking beams of light stretching far out from the nebula may be either the result of light escaping through holes in the shell surrounding the star, or light from jets of matter being ejected from both poles of the central star.

—— PLATE 29 ——

HOURGLASS NEBULA

The planetary nebula surrounding the star MyCn 18 is 8000 light years distant. The hourglass shape of the gas expelled from MyCn 18 is due to the gas from the poles being less dense than that at the equator. This gas moves faster out from the star, leaving the characteristic shape of the nebula. The eye shape at the centre of the nebula may be the result of a very recent ejection of mass which will grow to be like the hourglass.

Observations have shown that MyCn 18 does not lie at the exact centre of the Hourglass Nebula. This suggests that it may have a companion star which cannot be seen and which may be acting to push gas further out at the poles of the nebula, enhancing the hourglass shape.

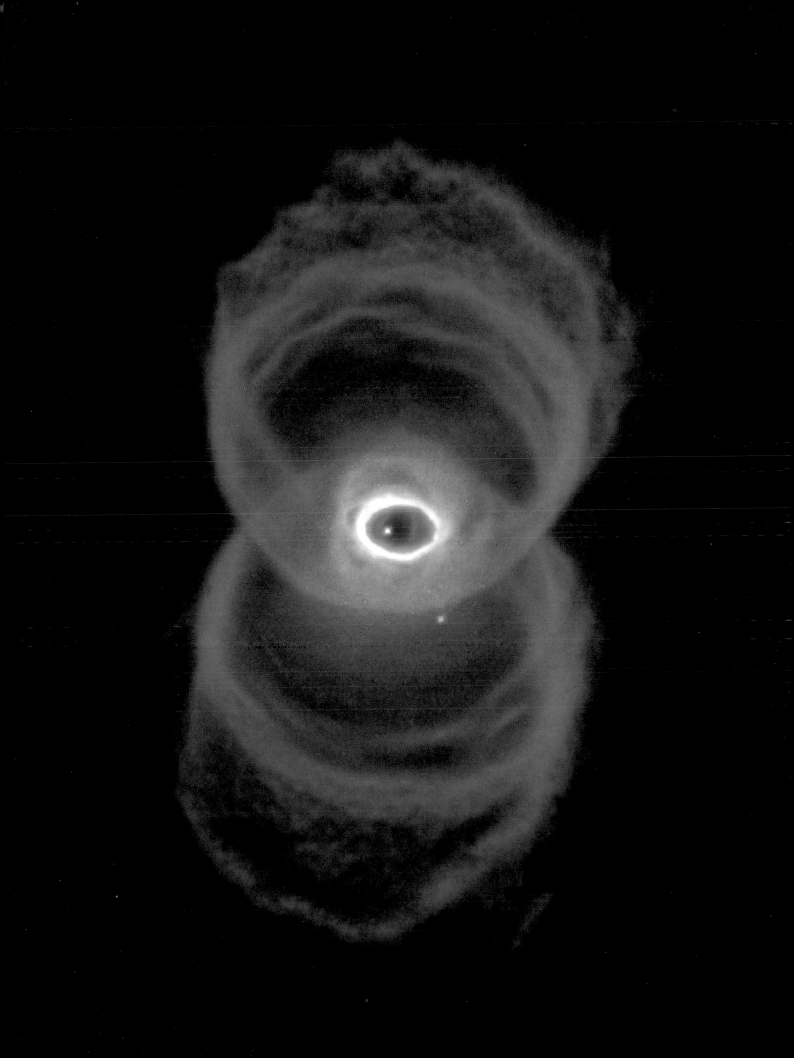

—— PLATE 30 ——

CAT'S EYE NEBULA (NGC 6543)

The Cat's Eye Nebula (NGC 6543) is one of the most complex planetary nebulae. It is 3000 light years away in the constellation Draco, and its intricate structure is probably associated with the existence of a close companion star to the central red giant. The two stars are too close to be resolved as separate objects, even by the HST. The appearance of the nebula is due to several different phases of gas loss from the red giant. Much of this loss was in the form of rings of material shed in the plane of the orbit of the smaller star around the red giant.

The colours in this composite WF/PC2 picture show the types of atom that are radiating in various parts of the nebula, red indicating hydrogen, blue oxygen and green nitrogen. The green arcs of glowing nitrogen are possibly produced by two jets shooting out from the companion star. Material falling onto this star is accelerated and forced along the polar axis of the star at high speed. As in the Herbig-Haro jets (plate 23), this high speed gas slows down when it hits gas in the nebula and the interstellar medium at the edge of the nebula. This causes the gas to glow as it loses energy. The shape of these features suggests that the jet wobbles as it is shot out, spreading material over a large volume of the nebula.

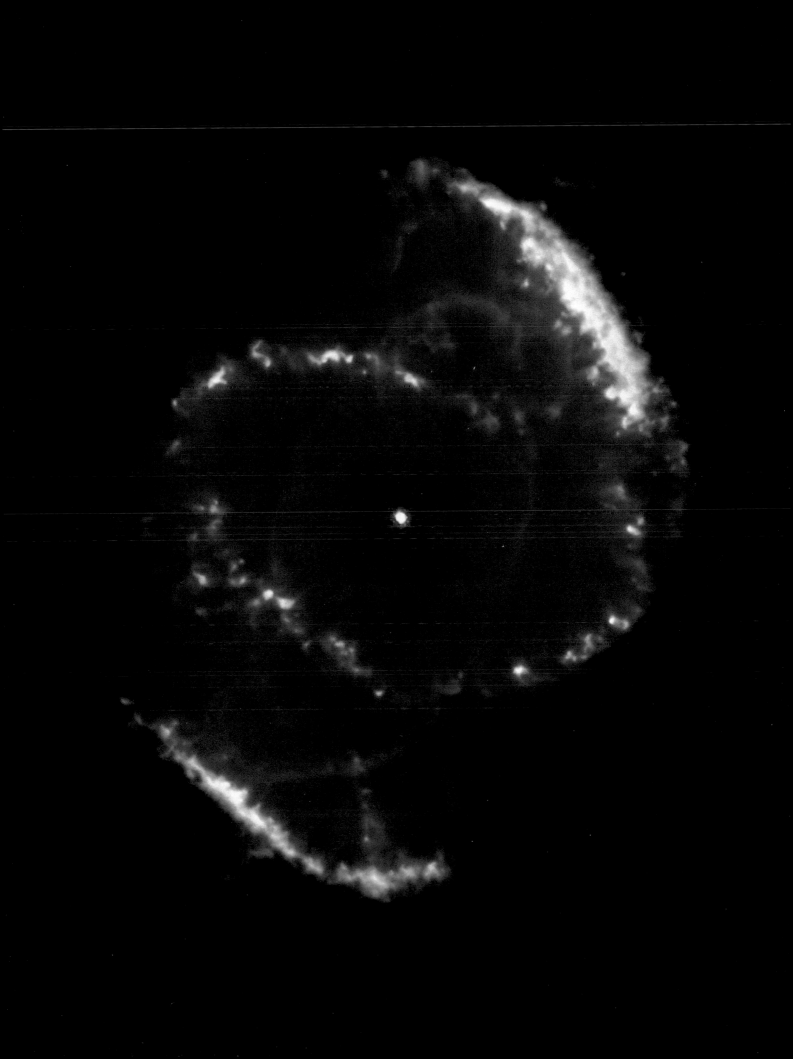

—— PLATE 31 ——

HELIX NEBULA

The Helix planetary nebula is the closest nebula to us, only 450 light years distant. It is so close that its apparent size on the sky is half that of a full Moon – very large by the standards of most astronomical objects – although it is very faint, and not easily visible to the naked eye.

This WF/PC2 picture shows a portion of the edge of the nebula, a few trillion km from the central star, revealing thousands of tadpole-shaped gaseous blobs called 'cometary knots'. The cometary knots, however, have no relationship to the comets in the Solar System. Each head is larger than the entire Solar System and the tails stretch for at least 200 billion km. It is thought that they are formed when a hot shell of gas ejected by the central star collides with a cooler, older shell ejected some time earlier. The mixing of hot and cold gas causes condensation of fingers of denser gas which appear to us as these cometary knots.

The knots will evaporate within a few hundred thousand years, but an interesting possibility is that dust in their centres might collapse to form Earth-sized objects similar to a cometary nucleus. These 'planets' would then escape to roam the Galaxy. If this were to happen, then interstellar space could contain trillions of these bodies formed by planetary nebulae.

In this picture the colours show what atoms are emitting light. Green is hydrogen, blue oxygen and red nitrogen.

—— PLATE 32 ——

SUPERNOVA 1987A

In February 1987, in the Large Magellanic Cloud, a star 20 to 25 times larger than the Sun was seen to explode in one of the rarest and most spectacular events: a supernova. In a supernova a huge star very violently ejects most of its mass in an event lasting only a few seconds. In this time the star creates almost ten times the amount of energy as the Sun will produce in its 10 billion year lifetime. Unsurprisingly, supernovae are extraordinarily bright: in small galaxies a supernova can rival the brightness of the whole galaxy. Supernova 1987A is the closest supernova to the Earth since there have been telescopes. In this millennium only five supernovae have been observed in our Galaxy, the most spectacular in 1054.

There are two types of supernova, and 1987A was an example of the more powerful: a type II supernova. When stars larger than about ten times the size of the Sun exhaust their supplies of helium at the end of their red giant phase, they switch to another type of fuel. Their cores contract, and are large and hot enough to burn carbon by nuclear fusion to produce heavier and heavier elements. Eventually, when iron is formed in the core, this burning stops producing energy. When this happens the core of the star will suddenly (in a tiny fraction of a second) collapse down to a few kilometres across. The energy released by this collapse pushes the outer layers of the star away at speeds of up to 10 000 k.p.s. The superheated shell of gas ejected from the supernova expands into space, and it is this that is observed as the supernova.

WF/PC2 took this picture in visible light two years after the supernova to investigate what had happened since. The picture shows the gas ejected by supernova 1987A when it exploded. Contrary to what astronomers were expecting, the two large rings are in front of and behind the place where the supernova occurred. It is thought that they are parts of the shell of the supernova that are being illuminated by jets of high-energy radiation from a previously unknown neutron star that was a companion of the star that exploded. The two bright, white points of light are other stars in the field of view.

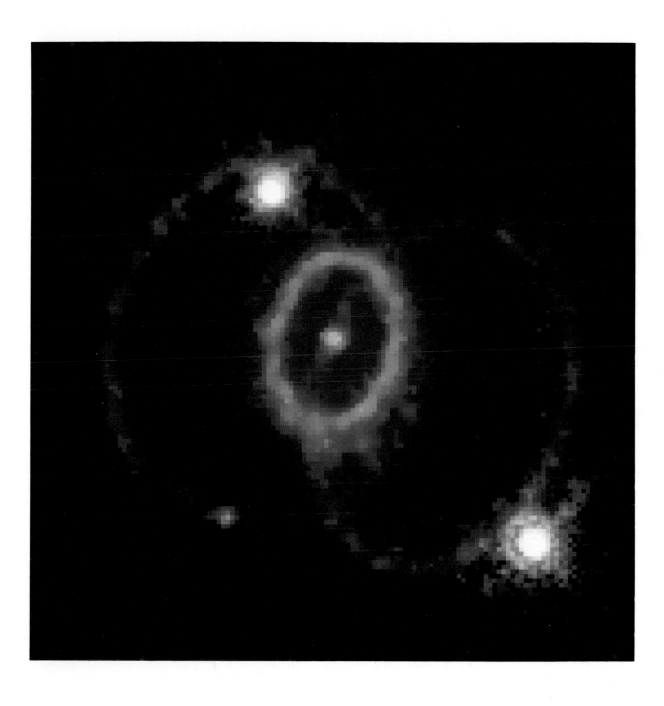

—— PLATE 33 ——

CYGNUS LOOP (SUPERNOVA REMNANT)

When a supernova explodes, it sends large amounts of gas into space at extremely high speeds. The rapidly expanding shell creates a shock wave where it meets interstellar gas, superheating this gas to millions of degrees. The hot gas emits light, marking the progress of the shell as it continues to expand, gradually thinning out, for many thousands of years. The most spectacular example of such a shell is the Crab Nebula left after the supernova of 1054.

The shell of gas expands outwards into the interstellar medium, a very tenuous gas that fills most of the void between the stars. The interstellar medium normally contains only a few atoms in every cubic centimetre. As the shock wave from the supernova sweeps up the interstellar medium, it leaves a hole in the medium and gains mass as it passes. As the shock wave expands it slows when hitting the gas in the interstellar medium. These collisions cause the gas to radiate energy and glow.

The catastrophic collapse of the core of the star creates such high temperatures and pressures that many nuclear reactions occur in the gas shell surrounding it in the first second or so. These reactions produce neutrons that build up very heavy atoms by colliding with other atoms. Supernovae are the only events in the Universe that can create elements heavier than iron. Eventually these heavy elements will end up in a nebula where they will be included in the formation of a new star and its planets. All of the heavy elements now present on Earth were created in this way, in the exploding heart of a huge, dying star.

This WF/PC1 pre-servicing mission image shows a small portion of the huge supernova remnant, 2600 light years away, known as the Cygnus Loop. The entire Cygnus Loop is 3° across on the sky, six times the extent of the full Moon. Probably around 15 000 years old, the shock wave has been slowed by the interstellar medium to a respectable 100 k.p.s. The picture is a composite of images in three colours, each corresponding to the radiation from one type of atom. Oxygen, at a temperature of 30 000 to 60 000°C, shows blue; sulphur, in cooler regions of about 10 000°C, shows red. The yellow colour represents hydrogen which is present throughout the shock wave.

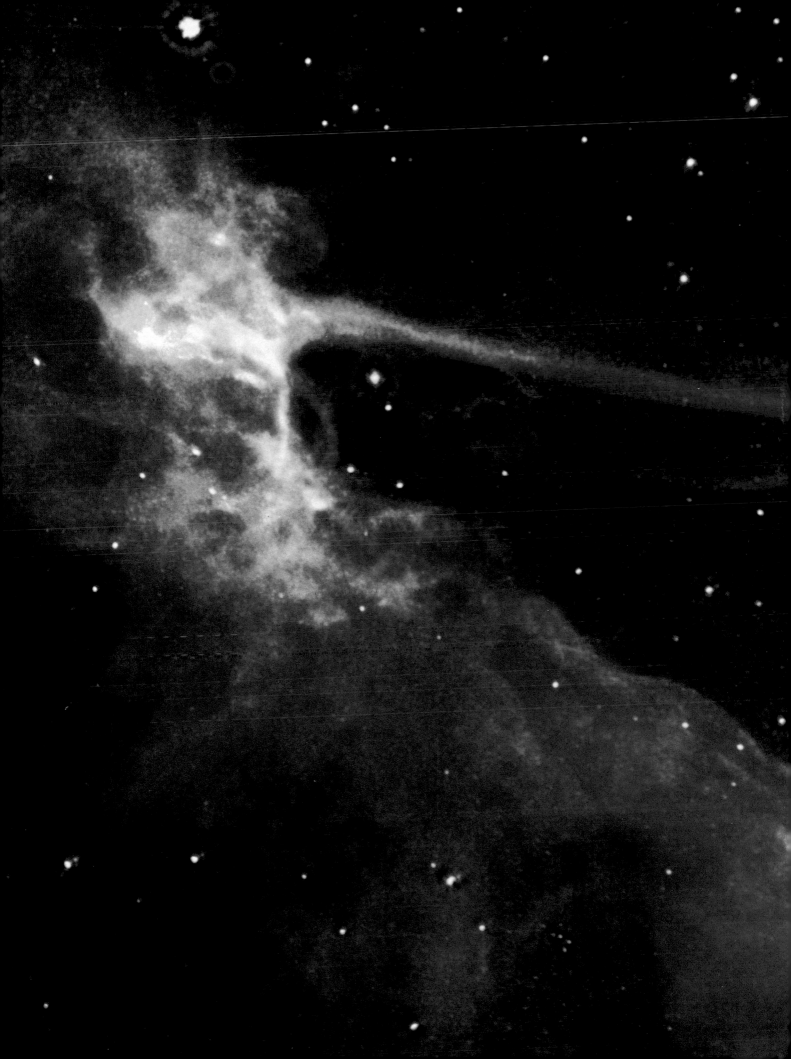

—— PLATE 34 ——

SEARCH FOR DARK MATTER IN STAR CLUSTER NGC 6397

Stars in the Milky Way move under the gravitational influence of all the other objects in it. If the motions of stars are studied it is possible to work out how heavy the Milky Way is. The method is the same as that used to calculate the masses of the Earth, Moon, Sun and other objects in the Solar System. When this calculation is made for the Milky Way, it is found that the amount of matter in the Galaxy must be more than is contained in stars, gas and other material whose radiation at visible and other wavelengths we can detect. When these calculations are made for clusters of galaxies it is found that maybe less than half the amount of mass in the cluster can be seen; in some clusters perhaps as little as 10%. This is the problem of what is termed missing mass, or 'dark matter'. We call this material 'dark' because we cannot see it.

As this matter cannot be seen, we do not know what it is. It may be in the form of brown dwarfs, so far away that we cannot detect the tiny amounts of infrared radiation they emit. Others think that it may be exotic particles formed in the Big Bang that are virtually impossible to detect except by their gravitational pull on other particles.

Whatever it may be, the existence of this dark matter is tied up with the future of the Universe. If the Universe contains more than a particular amount of mass, known as the 'critical density', then at some time in the future it will collapse back on itself. If, on the other hand, it contains less than that amount of mass, then it will continue to expand for ever. Many theories of cosmology predict that the Universe should have just the critical density of matter, but observations appear to disagree. Matter that we can see makes up about 20% of the critical density, but nobody knows how much dark matter the Universe contains.

These WF/PC2 pictures were taken as part of a search for dark matter in the Milky Way. They show the central regions of the globular cluster NGC 6397, 7200 light years away. The HST was looking to see if the dark matter in our Galaxy might be in the form of very low mass stars and brown dwarfs. If the dark matter is in these tiny, dim stars then the HST would be able to see them, something that has never been possible from the surface of the Earth. Observations were made of NGC 6397, and calculations were made of what the HST should be able to see if there were lots of tiny stars in the Milky Way. When they were compared, it was found that there are not enough stars to make up the weight of the Milky Way. Had there been enough, then the HST would have seen an extra 300 stars (one for each of the diamonds in the lower picture). As it was, the HST saw only 200 stars.

These observations suggest that most of the Universe could be made up of very strange, unknown particles, and that what we see is only a tiny fraction of what the Universe actually is.

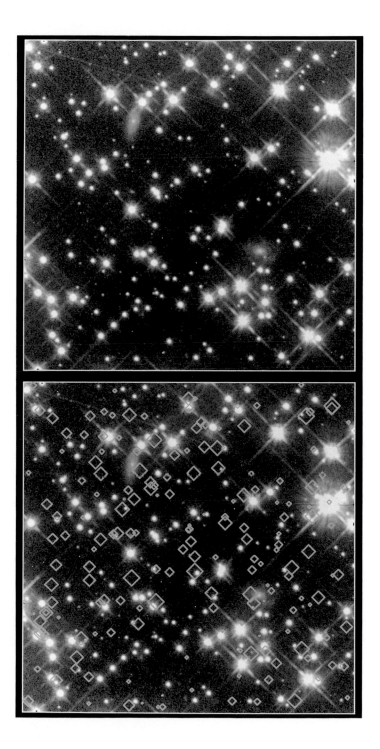

—— PLATE 35 ——

SPIRAL GALAXY M100

M100 is a spiral galaxy in the Virgo Cluster, which is the nearest large cluster of galaxies and contains many spectacular examples of different types of galaxy. Spiral galaxies, of which the Milky Way is one, consist of a thin disc of stars and gas spinning around a central bulge of stars. The spiral arms that are the most obvious feature of these galaxies are huge bands of star formation. There are actually no more stars in the arms than anywhere else in the disc of a spiral galaxy. It is just that the arms contain a lot of massive, young stars that shine brightly and make the arms stand out against the background of stars in the disc, hence their bluer colour. The star formation in these arms is created by a 'density wave' that travels around the disc and create the spiral pattern. Normally spiral galaxies have two prominent arms, but one or three have been seen in some galaxies.

This WF/PC2 image of M100 shows the galaxy in unprecedented detail. It is possible to observe individual stars within the galaxy, where previous images had been too blurred.

—— PLATE 36 ——

CEPHEID VARIABLE IN M100

The reason that the HST was observing the spiral galaxy M100 (plate 35) was to look for a type of star known as a Cepheid variable. These are stars which change their brightness on a regular time scale measured in days. The time scale on which they vary is linked to their intrinsic brightness. If a Cepheid variable can be observed over a period of time to find its period of variation, then its actual brightness can be calculated, hence also its distance from Earth. If Cepheid variables can be seen in other galaxies, then we can then find the distance to these galaxies. Cepheid variables are the most important of the 'standard candles'.

The HST was able to find 20 Cepheid variables in M100. In the pictures shown here, this Cepheid can clearly be seen to dim between 23 April and 9 May and brighten again thereafter. From this information it can be calculated that M100 is 56 million light years (plus or minus 6 million light years) away. This distance gives an idea of how far away other galaxies in the Virgo Cluster are.

The redshift of M100 is also known, and so it should be possible to work out the Hubble Constant. M100 is moving not just with the speed of the expansion of the Universe but also under the gravitational influence of all the other galaxies in the Virgo Cluster. M100 is not far enough away for the 'peculiar velocity' caused by the other galaxies around it to be small in comparison with the expansion of the Universe. Cepheids in other galaxies, far more distant, will have to be found before the size and age of the Universe can be calculated in this way.

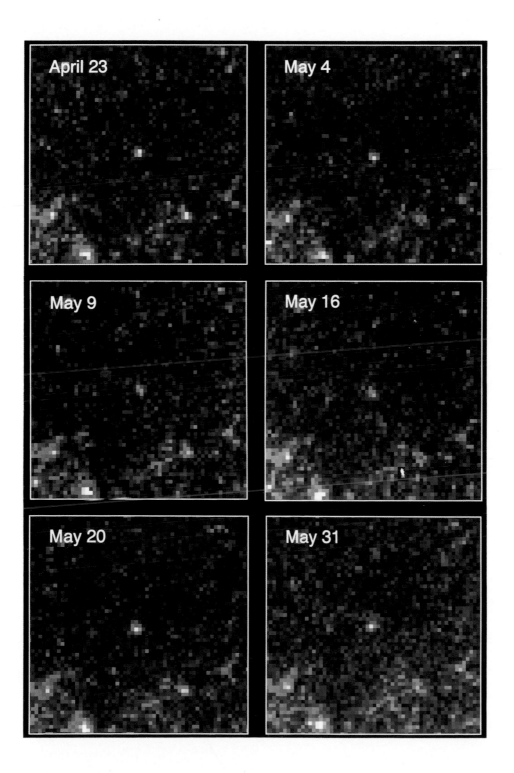

April 23

May 4

May 9

May 16

May 20

May 31

—— PLATE 37 ——

STARBURST GALAXY NGC 253

Starburst galaxies are galaxies in which a large amount of star formation is occurring, often in a fairly small region of space. They radiate far more strongly in the infrared than in the visible part of the spectrum. This infrared radiation is caused by the large amounts of dust and gas associated with the star formation. The dust in the clouds absorbs ultraviolet radiation from the new stars and re-radiates it as lower energy infrared light. Starburst galaxies are normally spiral galaxies, as they contain enough gas to produce so many new stars. The cause of starbursts remains unknown. Sometimes a neighbouring galaxy may have initiated the star formation by its tidal pull on the starburst galaxy. Yet other starbursts have no near neighbours, so another process must be at work as well.

This WF/PC2 picture shows the central 1000 light years of NGC 253, 8 million light years away in Sculptor, in unprecedented detail. Within this region stars appear to be forming in several compact regions (bright white areas), with lanes of dust and gas crisscrossing them.

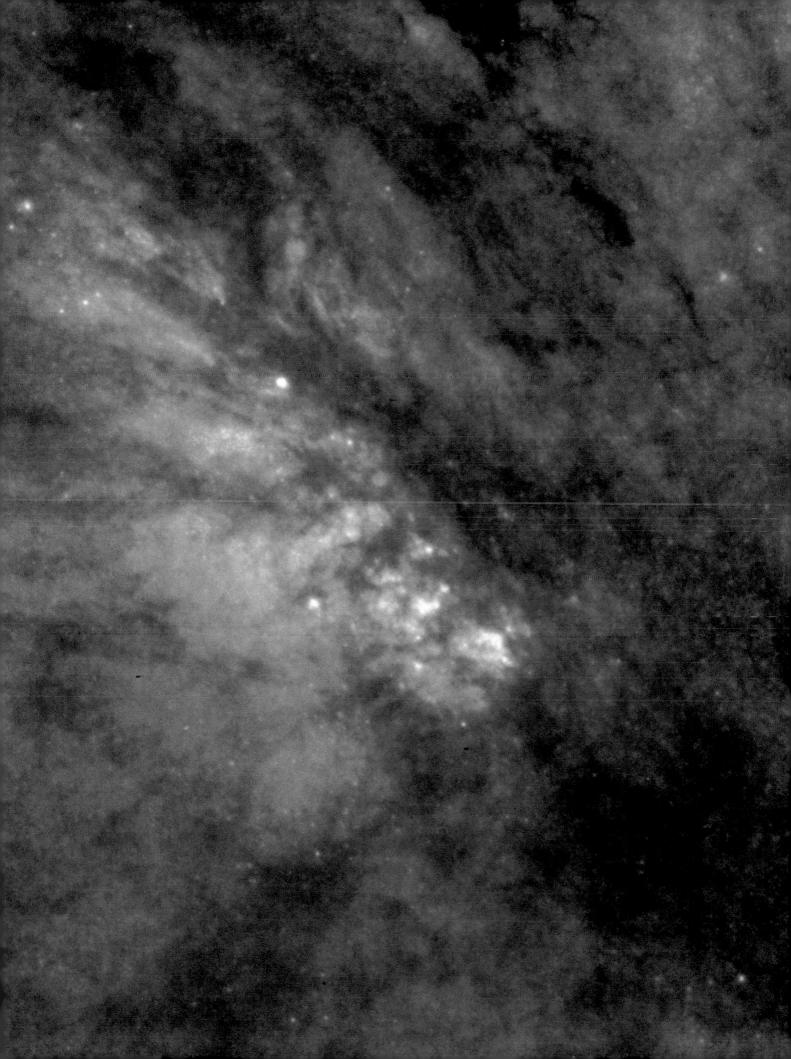

—— PLATE 38 ——

CARTWHEEL GALAXY

The Cartwheel Galaxy is a large galaxy 500 million light years distant in the constellation of Sculptor. The bright, blue ring in this galaxy shows that a very unusual event has occurred here. The ring is 150 000 light years in diameter and made of bright young stars. The huge amounts of star formation in the ring were probably caused by a massive shock wave speeding outwards at about 300 000 k.p.h. from the centre of the galaxy. This shock wave was probably produced by another galaxy ploughing through the larger galaxy at high speed. Before the collision, the Cartwheel Galaxy probably appeared very much like our own Milky Way spiral galaxy. The 'spokes' of the Cartwheel are probably the original spiral arms, faint against the far brighter ring. As the shock wave passed through the galaxy it will have caused star formation throughout the galaxy. As the bright stars inside the ring have aged and died, they have faded. This means that we see clearly only the most recent areas of star formation.

The picture was taken by WF/PC2, and is a composite of images in blue and infrared light. It shows, in remarkable detail, the structure of the ring. Stars have formed in clusters as the shock wave has passed. This gives the ring its slightly lumpy structure.

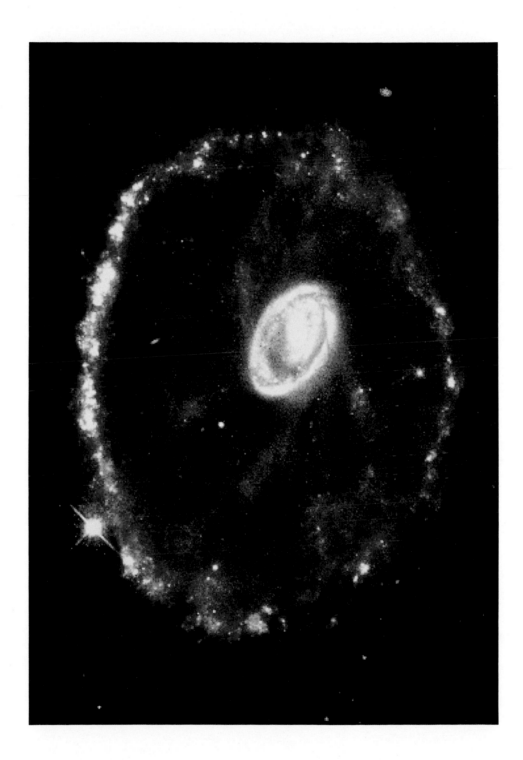

—— PLATE 39 ——

CARTWHEEL GALAXY (DETAIL)

There are only two galaxies near the Cartwheel sufficiently close that they could have recently passed through it and caused a shock wave to travel through the galaxy. These two galaxies are far smaller and present something of a mystery to astronomers. Neither have quite the appearance that would be expected if they had recently passed through a much larger galaxy.

The upper galaxy is very smooth and undisturbed. If it had recently passed through the Cartwheel, it would be expected to have an irregular appearance as material from it would have been ripped out of it by the gravitational pull of the much larger galaxy.

The lower of the two galaxies looks as if it has recently been disturbed by a close encounter. It would seem to be the obvious candidate, except that it still contains gas which we would expect to have been stripped out of it by any passage through the Cartwheel Galaxy. This gas itself cannot be seen, but the bright blue colour of the galaxy shows that there are a lot of young stars within it that could not have formed if there were no gas in the galaxy.

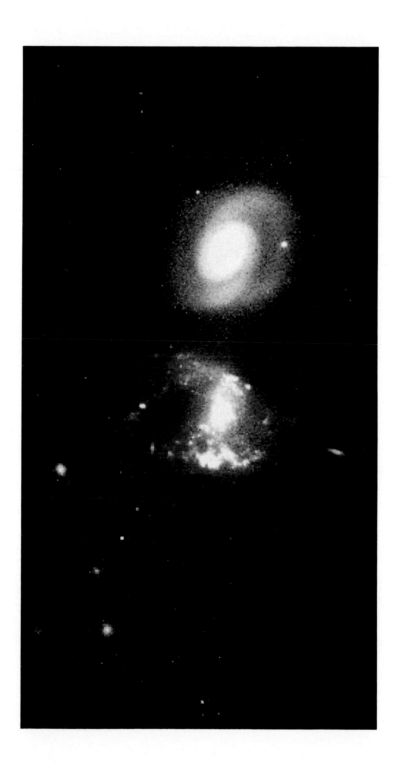

—— PLATE 40 ——

GALAXY NGC 4881 IN THE COMA CLUSTER

NGC 4881 is a massive elliptical galaxy near the edge of the Coma Cluster of galaxies. Unlike spiral galaxies, elliptical ones have exhausted their supplies of gas and no longer form stars. The Coma Cluster contains at least 1000 bright galaxies, both spiral and elliptical; these galaxies all lie within a sphere approximately 20 million light years across.

In the background of this WF/PC2 picture a large number of other galaxies, most of them in the Coma Cluster, can be seen. At the top right is a fine example of a face-on spiral galaxy. Around NGC 4881, as around any large galaxy, are a number of globular clusters (see plates 22 and 34). The HST was looking for these in order to try to estimate the distance to NGC 4881. Globular clusters always appear to have the same mean brightness, no matter what galaxy they orbit. By measuring the mean brightness of the globular clusters around NGC 4881 and comparing this with other galaxies of known distance, NGC 4881 was calculated to be more than 300 million light years (just under 100 million parsecs) away. When the redshift of NGC 4881 is measured it is found to be moving away from us at about 7000 k.p.s.; the Hubble Constant is therefore estimated to have a value of 70 k.p.s. per million parsecs.

—— PLATE 41 ——

HUBBLE DEEP FIELD SURVEY

The Hubble Deep Field Survey was used to study galaxies further away than ever before. The images were taken over a period of 150 orbits of the HST in the direction of the North Galactic Pole. This direction was chosen as it points straight up out of the disc of the Milky Way and into intergalactic space, with the minimum possible amount of gas and dust to obscure the view. Each of the 342 exposures made by the HST lasted 15 to 40 minutes and was taken in four colours from the infrared to blue light. Using these four colours it is possible to learn much about all the galaxies in the images. The long exposures enabled as much light as possible to enter the telescope. Some objects in the picture are so faint that they have never been seen before, and may be nearly as old as the Universe itself. These faint galaxies are 4 billion times fainter than the human eye can see.

The view provided by the Deep Field Survey covers only a tiny portion of the sky, only one-thirtieth of the diameter of the full Moon, and is thought to be a representative sample of the Universe in any direction. The picture contains at least 1500 images that are identifiable as separate galaxies, of all types. The images from the survey were released to the astronomical community as soon as they were available, in mid-January 1996. These data are in the very early stages of analysis, and many discoveries are expected.

—— PLATE 42 ——

FAINT BLUE GALAXIES

The Medium Deep Survey of 1995 (see also plate 43) was intended to shed light upon the 'faint blue galaxy mystery'. Carried out over 48 orbits of the HST, it covered an area of sky nine times larger than the Deep Field Survey (see plate 41) but in slightly less detail. The survey found that the most common type of galaxy in the Universe of around 4 to 8 billion years ago looks faint and emits more light in the blue part than in other parts of the visible spectrum. However, at the present time this type of galaxy seems to have disappeared. What happened to them?

One idea is that they are blue due to star formation, and it is bright young stars that make them visible. In the Universe today they have faded as stars within them have died and have become 'invisible', rather than having disappeared. Or maybe the combined effects of many supernovae in the galaxies blew them apart, spreading the stars through intergalactic space. Another possibility is that they have merged together over time. What were lots of very small galaxies have become fewer, larger galaxies.

—— PLATE 43 ——

DISTANT IRREGULAR GALAXIES

When the Medium Deep Survey (see plate 42) was analysed, a large number of faint blue galaxies were, indeed, present in the images. These pictures show what these galaxies looked like several billion years ago. Ground-based telescopes had previously resolved these galaxies as featureless blobs. They all appear to belong to a class known as irregular galaxies on account of their lack of distinctive shape. Like spiral galaxies they contain a lot of gas and dust, and are often involved in star formation. All these galaxies contain bright blue patches that indicate virulent star formation.

These images have not solved the mystery of the faint blue galaxies. Rather, they have raised more questions. Why are these galaxies so distorted? Why are they undergoing star formation at this time when larger galaxies, both elliptical and spiral, formed many or all of their stars far earlier in the history of the Universe?

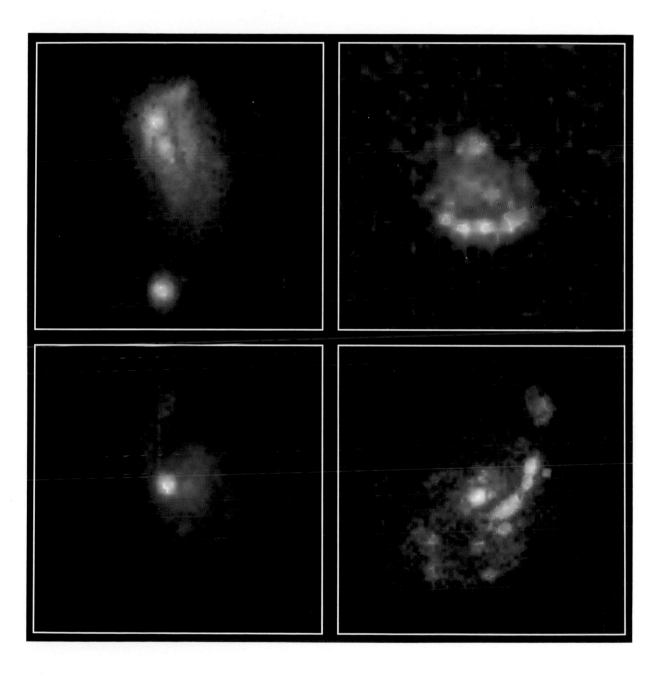

—— PLATE 44 ——

EVOLUTION OF GALAXIES

After the Big Bang, the Universe was a smooth, nearly uniform mixture of hydrogen and helium gas. After a billion years or so, the Universe had become 'clumpy'. Under the action of gravity lumps of gas had grown and were able to form stars. We see these masses of gas today as galaxies.

This group of pictures can be regarded as a series of snapshots in time, showing how the two basic types of galaxy – elliptical (e.g. NGC 4881, plate 40) and spiral (e.g. M100, plate 35) – have evolved. The top row shows galaxies as we see them today, some 10 to 20 billion years after the Big Bang. Further down the page the galaxies become more distant, hence older, back to not long after the beginning of the Universe itself.

The far column of elliptical galaxies shows that they have remained comparatively unchanged with time. The images of spiral galaxies, however, show considerable change: those of a few billion years ago are far more diffuse than today, indicating that they were still settling into their present structures. As the HST probes even deeper into time, the distinctive spiral structure becomes vague. Soon after the Big Bang the distinction between spirals and ellipticals is difficult to make. This evolutionary sequence is thought to be due to the constant star formation during a spiral galaxy's life.

—— PLATE 45 ——

QUASAR PKS 2349

When quasars were first discovered in 1963 they looked like bright pin-points of light, much like stars. For this reason they were first called quasi-stellar objects (QSOs), or quasars. The astronomer Maarten Schmidt realised that the spectra of quasars have been redshifted by the expansion of the Universe far more than any spectra seen before, meaning that they are further away than anything ever seen before. Many quasars are so far away that they appear to be moving away from us at over 90% of the speed of light (over 270 000 k.p.s.), putting them over 10 billion light years distant.

This realisation raised far more questions than it answered. If quasars were so far away then they must be incredibly bright for us to see them. In fact, they would have to be brighter than 100 normal giant galaxies put together. However, observations showed that quasars were very small, maybe only a few light years across at most. This was known because quasars varied their brightness on this time scale. The only thing that would be able to provide this much energy in such a small volume of space would have to be a super-massive black hole, weighing more than a billion times more than the Sun.

This picture shows the quasar PKS 2349 as the bright point. Faintly visible around the quasar is the galaxy in which the black hole is embedded. This galaxy looks as if it is being disrupted by merging with another galaxy. This result was unexpected as before it was thought that quasars could only exist in spiral galaxies that were not being disturbed. Around it are other galaxies that are presumably part of a galaxy cluster of which PKS 2349's host galaxy is a member.

—— PLATE 46 ——

BLACK HOLE IN M87

M87 is a huge elliptical galaxy 50 million light years away, the dominant, central galaxy of the Virgo Cluster of galaxies. For a long time it has been known that M87 is a very unusual galaxy, a member of a class known as active galaxies.

In the core of M87 is a radio source that first brought the galaxy to the attention of astronomers as being active. This radio source is only 1.5 light months in diameter, but radiates more radio energy than the entire Milky Way. In addition there is a jet, 6000 light years long, extending from the core. This jet shines brightly in optical wavelengths and is 25 million times brighter than the Sun. M87 also produces even more energy in X-rays than it does in visible light from the core and jet.

The production of the amount of radiation that M87 emits requires more than just stars. Some process is occurring at the very centre of M87 that is creating vast amounts of energy in a small volume of space (a matter of a few tens or hundreds of cubic light years). The only object known to science that could possibly produce this energy would be a black hole of the same size as those that power quasars.

This picture from WF/PC2 shows the centre of M87. At the bottom left is the position of the black hole. Around it is very energetic gas, heated up by the outflow of huge amounts of energy from around the black hole. The picture also shows a straight line across the middle. This is not a diffraction spike, but a jet of electrons, stripped from atoms and accelerated to 99% of the speed of light or more. One of these jets comes from near each of the two poles of the black hole.

—— PLATE 47 ——

BLACK HOLE IN M87 (2)

Black holes are formed when an object becomes so massive, or so small, that it can no longer withstand the pull of its own gravity. The object collapses to a mathematical point with no size, and then nothing that gets too close can escape its gravitational pull. Even light does not travel fast enough. It is possible that very massive black holes (with masses many million times that of the Sun) can form in the centres of most galaxies. It is even possible that the Milky Way contains a supermassive black hole. If conditions are right then a lot of gas, and even stars, will start to fall into this black hole. When they do they form an accretion disc, like those around young stars but far, far larger, possibly hundreds of light years across. As the material gets closer to the black hole it is heated up and gets so hot that it shines in X-rays and gamma rays. Electrons are speeded up to velocities close to that of light, and they spin around the magnetic field of the black hole giving out huge amounts of energy as radio waves. Some material is shot out of the disc and along the poles of the black hole at huge speeds, forming enormous jets that can be hundreds of thousands of light years long.

This detail of the core of M87 shows the disc around the black hole in M87. The black hole itself cannot be seen directly, as light cannot escape from it for us to see. But, by using the spectra of the gas in the disc 60 light years out from the centre, and measuring the red- and blueshifts of the lines, it is possible to find the speed of the disc. This gas is moving at up to 550 k.p.s. (about 2 million k.p.h.). Knowing this, the mass of the central object which is pulling in the gas can be estimated. In M87 this mass is three billion times that of the Sun. Nothing else that we know of in physics could be so small and so heavy – it must be a black hole.

—— PLATE 48 ——

BLACK HOLE IN NGC 4261

NGC 4261 is an active elliptical galaxy 100 million light years away, and is fuelled by a wide accretion disc of gas and dust at its centre. The disc contains the mass of 100 000 Suns spiralling towards a black hole 1.2 billion times heavier than the Sun contained in a volume the same as the Solar System's. Elliptical galaxies are expected to have very little gas and dust left in them, but this picture, like that of M87 (plate 46), clearly shows that there is a lot of gas present. This is presumed to have belonged to another, smaller galaxy that has fallen into NGC 4261 and has been 'eaten'. Many large elliptical galaxies are thought to have grown by 'cannibalising' other, smaller galaxies. Over the next 100 million years the black hole will use up all the gas and dust and will eventually run out of fuel; the galaxy will then cease to be active, leaving a normal elliptical once more. If this is the process that fuels many active galaxies and quasars, it would explain why there were so many more active galaxies and quasars in the early Universe, when these mergers were common, than there are today.

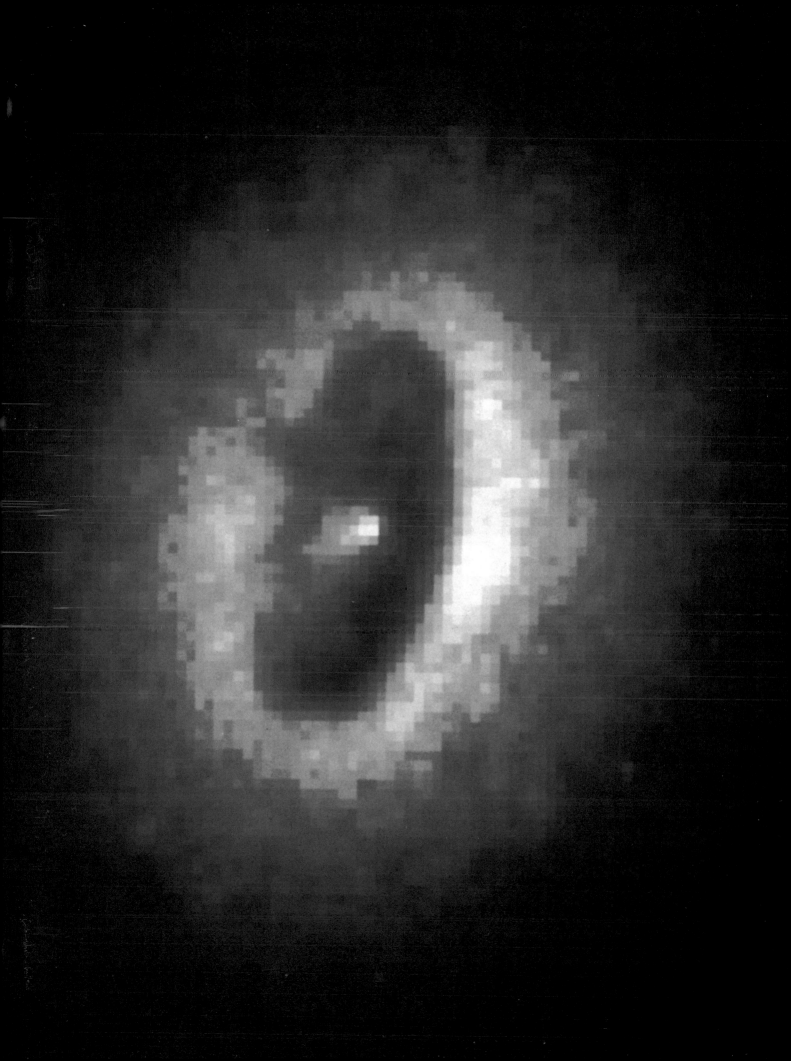

—— PLATE 50 ——

GRAVITATIONAL LENSES IN ABELL 2218

The richly populated galaxy cluster known as Abell 2218 is located in the constellation Draco, some 1 to 2 billion light years away. Everything that can be seen in this image is a galaxy. Most galaxies in clusters as massive as Abell 2218 are elliptical, especially in the centre, as the more fragile spiral galaxies are disrupted and warped by coming too close to other galaxies. There is also a huge amount of gas which is so hot (several million degrees Celsius) that it radiates very high energy X-rays, as well as large amounts of dark matter. The total mass of a galaxy cluster like this is at least 50 000 billion times that of the Sun.

The unusual arcs of light all around the cluster are the distorted, gravitationally lensed images of galaxies which are up to ten times more distant than Abell 2218 itself. In seven cases there are several images of the same distant galaxy; the rest are all single images, bent by the huge mass of Abell 2218. These lensed images are of galaxies so distant that they would be impossible to see if it were not for the gravitational pull of Abell 2218 which has brought them into focus for us. The spectra of these lensed galaxies can still be taken to find their redshifts and give information on what is happening inside the galaxy. This lens is in effect allowing us to look back to when the Universe was only a quarter of its present age.

GLOSSARY

Arcsecond Measure of angles often used in astronomy. One arcsecond is one 3600th of a degree, or less than one-millionth of the distance around a circle. The distance of a star which changes its apparent position by one arcsecond in six months is a parsec (PARallax SECond), about three-and-a-quarter light years. The WF/PC can see down to detail around one-tenth of an arcsecond.

Astronomical Unit (AU) The average distance between the Earth and the Sun, about 150 million km. The AU is a useful distance to describe the distances between planets.

Black hole An object that has been shrunk by gravity until its gravitational pull is so strong that nothing, not even light, can travel fast enough to escape from it. See plates 46, 47 and 48.

Brown dwarf An object that is like a star, but too small to become hot enough to burn hydrogen by nuclear fusion in its centre. Brown dwarfs never shine like stars but may be seen in the infrared. See plate 25.

Cepheid variables A type of very bright star which pulses with a regular period, thus regularly changing in phases of brightness and dimness. If the period of the pulses is known then the distance to these stars can be calculated very accurately. They are thus very useful in finding the distance to galaxies. See plate 36.

Charge coupled device (CCD) An electronic photographic plate used to take images in the HST. Tiny electronic eyes known as pixels receive light and turn it into a digital signal which can be stored on a computer. The CCDs in the WF/PCs are each 800 by 800 pixels and cover an area of the sky of about 8 square arcminutes, a part of the sky equivalent to 1 square metre on the surface of the Earth.

Comet A small body in the Solar System made up of lumps of ice and rock ('dirty snow-balls'). They become visible as they near the Sun, spewing out a tail made up of material boiled by the heat of the Sun. It is estimated that a comet can round the Sun about 1000 times before boiling away. See plates 13 and 14.

Corrective Optics Space Telescope Axial Replacement (COSTAR) COSTAR was installed in the repair mission in order to correct the fault in HST's main mirror. It corrected the paths of light rays from the main mirror to remove the fault.

Dark matter The name given to any matter that cannot be seen. As we cannot see it we do not know what it is, but we do know that it is there from the gravitational pull on objects we can see. As much as 90% of the Universe may be made up of mysterious dark matter. See plate 32.

Diffraction spike An optical flaw in an image caused by bright stars being overexposed in order to focus on fainter objects. Several pictures in this book have diffraction spikes (e.g. plate 25).

Faint Object Camera (FOC) A central instrument in the HST, designed to take images over a small area down to an extraordinary resolution of 0.06 arcseconds.

Faint Object Spectrograph The HST instrument designed to take the spectra of objects often too faint to be seen from the ground.

Fine Guidance Sensors (FGSs) The instruments that position the HST and make sure that it is constantly pointing towards the object the HST is observing. They are also used to find extraordinarily accurate distances to closer stars.

Galaxy A collection of hundreds of billions of stars, and often gas, held together by gravity. Spiral galaxies, of which the Milky Way is one, are characterised by their brightly shining arms of star formation (see plate 35). Elliptical galaxies (e.g. plate 40) do not contain gas and therefore cannot form new stars. Galaxies of less distinctive shape are known as irregular galaxies (e.g. plate 43) and appear to have been common in the early history of the Universe, which the HST is continuing to probe (see plates 41–44). Massive groups of individual galaxies bound together by the force of gravity are known as galaxy clusters (see plate 49).

Goddard High Resolution Spectrograph (GHRS) The GHRS takes the spectra of objects in the ultraviolet part of the spectrum which cannot be observed from the Earth.

Gravitational lenses When light from a distant galaxy or quasar is bent by the gravitational pull of another galaxy or galaxy cluster we can sometimes see multiple images of the distant galaxy. This effect is gravitational lensing. See plates 49 and 50.

H II region A cloud of gas made up of ionised hydrogen, for example the Tarantula Nebula (see plates 19 and 20). Normally hydrogen atoms have one proton being orbited by one electron. When hydrogen is ionised the electron is lost from the proton.

High Speed Photometer (HSP) The HSP was designed to look at particular frequencies and provide detailed information on how the light from a particular object changes with time. The HSP was affected by a number of the HST's teething problems and was replaced on the repair mission with COSTAR.

Hubble's Law In 1929 Edwin Hubble established his law of the expanding Universe, according to which the speed at which galaxies are moving away is proportional to their distance from Earth. The fundamental cosmological quantity known as the Hubble Constant is derived from this law and is a measure of how fast galaxies appear to be moving away from us with distance; at present this appears to have a value between 50 and 80 kilometres per second per million parsecs' distance.

Infrared light The part of the spectrum with frequencies just too long for the human eye to see.

Light year The distance travelled by light in one year. Light travels at 300 000 kilometres per second, so in one year it covers 9500 billion km, fast enough to circle the Earth seven and a half times in a second. Occasionally distances are measured in 'light days' or 'light months'.

M number The French astronomer Charles Messier (1730–1817) compiled a catalogue of 'nebulous' objects in the sky. Many of these are visible with the naked eye and are still commonly known by their number in Messier's catalogue, for example the Eagle Nebula (see plate 17).

Nebula A huge cloud of dust and gas, often weighing many millions of times more than the Sun. Nebulae are thought to be the birth places of virtually all stars, including the Sun.

NGC numbers The NGC number is the classification number from the *New General Catalogue of Nebulae and Clusters of Stars*, first published in 1888. Many objects in the sky are known by their NGC number.

Parsec See **Arcsecond**

Planetary nebula At the end of a star's red giant phase the outer layers of the star are blown away creating a shell of dust and gas around the star known as a planetary nebula (which have nothing to do with planets, although they were once thought to, hence their name). See plates 27–30.

Quasars Distant galaxies which give out incredible amounts of energy. This energy is thought to be produced in the centres of these galaxies around huge black holes which may be many millions of times heavier than the Sun. See plate 45.

Red dwarf A generic name given to any star less massive than the Sun. These stars are also cooler than the Sun and so appear more red, hence their name. Most stars belong to this category.

Red giant When a star has burnt all of the hydrogen in its core the star expands as it starts to burn helium. This expansion cools the surface of the star and it becomes a red giant.

Redshift In light from distant galaxies, the lines of the spectrum are of a lower frequency and therefore seem to have moved towards the red end of the spectrum. Hubble discovered that in the expanding Universe the redshift is proportional to distance. If the redshift of an object is known, its distance away from us can be calculated.

Spectra When light is split up into its component colours some thin bands of colour are found to be missing. These missing lines provide a great deal of information on the composition of an astronomical object which could not be found in any other way. The spectrum of a star, for instance, can tell us what atoms are present in the star and how many of them there are. Spectrographs, such as the Faint Object Spectrograph and the Goddard High Resolution Spectrograph on the HST, use prisms to split up the light and record what it looks like.

Standard candle The name given to any object whose actual brightness is known. Seeing how bright these objects appear to us when we see them in distant galaxies allows us to calculate how far away the galaxy containing them is. The best standard candles are Cepheid variable stars, but other sorts of stars as well as supernovae and planetary nebulae can be used as standard candles.

Star A mass of mostly hydrogen and helium gas which collapses together under gravity. As a star collapses it gets hot enough in the centre to begin the nuclear fusion of hydrogen into helium. This reaction creates huge amounts of energy and light which we are able to see.

Supernova When a star more than about ten times bigger than the Sun has reached the end of its life, it violently ejects its outer layers in an explosion which can be seen from millions of light years away. See plates 32 and 33.

Ultraviolet light The part of the spectrum of light with frequencies just too small for the human eye to see. The atmosphere blocks out the vast majority of the ultraviolet light from space, especially that from the Sun which can be extremely harmful.

White dwarf The core of a star that remains after the outer layers have been ejected during the red giant phase of a star's life. White dwarfs are no longer undergoing nuclear fusion and so cannot hold themselves up against the pull of gravity by producing this energy. They collapse down until they have a diameter only a little larger than that of the Earth. Over billions of years the heat stored in a white dwarf radiates away into space until all that is left is a tiny, cold black dwarf which is virtually undetectable. This will be the final fate of the Sun in around 5 billion years.

Wide-field/Planetary Camera (WF/PC) This equipment on the HST comprises three wide-field cameras and a smaller, more sensitive, planetary camera. Its four CCDs can image a wide angle in the sky in great detail; the CCD with a narrower field of vision is placed just off-centre, and this accounts for the chevron shape of some of the images produced. The original WF/PC1 was replaced during the repair mission with the WF/PC2 at the same time that COSTAR was installed. The WF/PC2 contained extra optics to counter the flaw in the HST's mirror.

ACKNOWLEDGEMENTS

Plates 1–6: NASA

Plate 7: Philip James (University of Toledo), Steven Lee (University of Colorado) and NASA

Plate 8: Reta Beebe, Amy Simon (New Mexico State University) and NASA

Plate 9: K. Noll (STScI), J. Spencer (Lowell Observatory) and NASA

Plate 10: J.T. Trauger (Jet Propulsion Laboratory), J.T. Clarke (University of Michigan), the WF/PC2 science team and NASA/ESA

Plate 11: Peter H. Smith (University of Arizona) and NASA

Plate 12: R. Albrecht (ESA/ESO Space Telescope European Coordinating Facility) and NASA/ESA

Plate 13: H.A. Weaver (Applied Research Corporation), P.D. Feldman (Johns Hopkins University) and NASA

Plate 14. H.A. Weaver and T.E. Smith (STScI) and NASA

Plate 15: H. Hammel (MIT) and NASA

Plate16: C. Robert O'Dell (Rice University) and NASA.

Plates 17 & 18: Jeff Hester and Paul Scowen (Arizona State University) and NASA

Plates 19 & 20: NASA

Plate 21: R. Gilmozzi (STScI/ESA), Shawn Ewald (Jet Propulsion Laboratory) and NASA

Plate 22: P. Guhathakurta (UCO/Lick Observatory, UC Santa Cruz), B. Yanny (Fermi National Accelerator Laboratory), D. Schneider (Pennsylvania State University), John Bahcall (Institute for Advanced Study) and NASA

Plate 23: Jeff Hestner (Arizona State University), WF/PC2 Investigation Definition Team and NASA

Plate 24: Mark McCaughrean (Max-Planck-Institute for Astronomy), C. Robert O'Dell (Rice University) and NASA

Plate 25: T. Nakajima (California Institute of Technology), S. Durrance (Johns Hopkins University) and NASA

Plate 26: J. Hestner (Arizona State University) and NASA

Plate 27: H. Bond (STScI) and NASA

Plates 28 & 29: Raghvendra Sahai and J.T. Trauger (Jet Propulsion Laboratory), the WF/PC2 science team and NASA

Plate 30: J.P. Harrington and K.J. Borkowski (University of Maryland) and NASA

Plate 31: C. Robert O'Dell and Kerry P. Handron (Rice University) and NASA

Plate 32: Christopher Burrows (ESA/STScI) and NASA/ESA

Plate 33: Jeff Hester (Arizona State University) and NASA

Plate 34: F. Paresce (STScI) and ESA

Plate 35: J.T. Trauger (Jet Propulsion Laboratory) and NASA

Plate 36: Wendy L. Freedman (Observatories of the Carnegie Institution of Washington) and NASA

Plate 37: Carnegie Institution of Washington and NASA

Plates 38 & 39: Kirk Borne (STScI) and NASA

Plate 40: WF/PC Team and NASA

Plate 41: Robert Williams and the Hubble Deep Field Team (STScI) and NASA

Plate 42: Rogier Windhorst and Simon Driver (Arizona State University), Bill Keel (University of Alabama) and NASA

Plate 43: Richard Griffiths (Johns Hopkins University), the Medium Deep Survey Team and NASA

Plate 44: A. Dressler (Carnegie Institution of Washington), M. Dickinson (STScI), D. Macchetto (ESA/STScI), M. Giavalisco (STScI) and NASA

Plate 45: John Bahcall (Institute for Advanced Study) and NASA

Plates 46 & 47: Holland Ford (STScI/Johns Hopkins University), Richard Harms (Applied Research Corporation), Zlatan Tsvetanov, Arthur Davidsen and Gerard Kriss (Johns Hopkins University), Ralph Bohlin and George Hartig (STScI), Linda Dressel and Ajay K. Kochbar (Applied Research Corporation), Bruce Margon (University of Washington) and NASA

Plate 48: L. Ferrarese (Johns Hopkins University) and NASA

Plate 49: Kavan Ratnatunga (Johns Hopkins University) and NASA

Plate 50: W. Couch (University of New South Wales), R. Ellis (Cambridge University) and NASA

INDEX

numbers refer to the Plate numbers